Planet Earth

SOLAR SYSTEM

TIME® LIFE BOOKS

Other Publications:

UNDERSTANDING COMPUTERS
YOUR HOME
THE ENCHANTED WORLD
THE KODAK LIBRARY OF CREATIVE PHOTOGRAPHY
GREAT MEALS IN MINUTES
THE CIVIL WAR
COLLECTOR'S LIBRARY OF THE CIVIL WAR
THE EPIC OF FLIGHT
THE GOOD COOK
THE SEAFARERS
WORLD WAR II
HOME REPAIR AND IMPROVEMENT
THE OLD WEST

For information on and a full description of any of
the Time-Life Books series listed above, please write:
    Reader Information
    Time-Life Books
    541 North Fairbanks Court
    Chicago, Illinois 60611

This volume is one of a series that examines the
wonders of the planet Earth, from its landforms,
seas and atmosphere to its place in the cosmos.

Cover
Like planets orbiting the sun, six of Saturn's
17 moons circle the huge ringed planet in this
montage of photographs taken by the *Voyager
1* spacecraft in November 1980. The montage,
symbolizing the dramatic discoveries made
by the Voyager missions, was assembled by the
National Aeronautics and Space Administration (NASA), which has been in the forefront of
space exploration since 1958.

Planet Earth

SOLAR SYSTEM

By Kendrick Frazier
and The Editors of Time-Life Books

Time-Life Books, Alexandria, Virginia

Time-Life Books Inc.
is a wholly owned subsidiary of

TIME INCORPORATED

FOUNDER: Henry R. Luce 1898-1967

Editor-in-Chief: Henry Anatole Grunwald
President: J. Richard Munro
Chairman of the Board: Ralph P. Davidson
Corporate Editor: Jason McManus
Group Vice President, Books: Reginald K. Brack Jr.
Vice President, Books: George Artandi

TIME-LIFE BOOKS INC.

EDITOR: George Constable
Executive Editor: George Daniels
Editorial General Manager: Neal Goff
Director of Design: Louis Klein
Editorial Board: Dale M. Brown, Roberta Conlan,
Ellen Phillips, Gerry Schremp, Donia Ann Steele,
Rosalind Stubenberg, Kit van Tulleken,
Henry Woodhead
Director of Research: Phyllis K. Wise
Director of Photography: John Conrad Weiser

PRESIDENT: William J. Henry
Senior Vice President: Christopher T. Linen
Vice Presidents: Stephen L. Bair, Edward Brash,
Robert A. Ellis, John M. Fahey Jr., Juanita T. James,
James L. Mercer, Wilhelm R. Saake, Paul R. Stewart,
Leopoldo Toralballa

PLANET EARTH

SERIES DIRECTOR: Gerald Simons
Picture Editor: Neil Kagan
Designer: Albert Sherman

Editorial Staff for *Solar System*
Text Editor: Russell B. Adams Jr.
Researchers: Jean Crawford, Martha Reichard George
(principals), Patti H. Cass, Roxie France, Sara Mark
Assistant Designer: Cynthia T. Richardson
Copy Coordinators: Elizabeth Graham,
Robert M. S. Somerville
Picture Coordinator: Renée DeSandies
Special Contributors: Karen Jensen, John Manners,
James I. Merritt, Charles C. Smith (text)

Editorial Operations
Design: Ellen Robling (assistant director)
Copy Chief: Diane Ullius
Editorial Operations: Caroline A. Boubin (manager)
Production: Celia Beattie
Quality Control: James J. Cox (director)
Library: Louise D. Forstall

Correspondents: Elisabeth Kraemer-Singh (Bonn);
Margot Hapgood, Dorothy Bacon (London); Miriam
Hsia (New York); Maria Vincenza Aloisi, Josephine
du Brusle (Paris); Ann Natanson (Rome). Valuable
assistance was also provided by: John Dunn
(Melbourne); Felix Rosenthal (Moscow); Christina
Lieberman (New York); Ann Wise (Rome); Dick
Berry (Tokyo); Traudl Lessing (Vienna); Bogdan
Turek (Warsaw).

Library of Congress Cataloguing in Publication Data
Frazier, Kendrick.
  Solar system.
  (Planet earth)
  Bibliography: p.
  1. Solar system. 2. Earth. I. Time-Life Books.
II. Title. III. Series.
QB501.F73 1985   523.2   84-16117
ISBN 0-8094-4529-8
ISBN 0-8094-4530-1 (lib. bdg.)

THE AUTHOR

Kendrick Frazier is a science writer and editor with lifelong interests in astronomy, planetary science and space exploration. He was editor of *Science News* from 1971 to 1977, and he has since been writing and editing as a freelancer. He is the author of two books, *The Violent Face of Nature* and *Our Turbulent Sun,* and numerous articles on scientific subjects for international publications.

THE CONSULTANTS

General Consultant
Stephen E. Dwornik is a longtime veteran of the National Aeronautics and Space Administration. He served as the Program Scientist on the Surveyor mission to the moon and on the *Mariner 10* mission to Venus and Mercury. Then, as Chief of Planetary Geology, he established the Comparative Planetology Program and the Planetary Cartography Program. He retired from NASA in 1979 and became the Washington, D.C., representative for an aerospace company, responsible for coordinating spacecraft and instrument programs for NASA space missions.

Consultants
Dr. Michael H. Carr is a geologist with the Branch of Astrogeology at the U.S. Geological Survey and a member of NASA's Voyager and Galileo Imaging Teams. He is the author of *The Surface of Mars* and the editor of the NASA publication *The Geology of the Terrestrial Planets.*

Geoffrey R. Chester is Production Coordinator at the National Air and Space Museum's Planetarium. A well-known lecturer on astronomy, he is the president of National Capital Astronomers.

Dr. Torrence V. Johnson is a Senior Research Scientist at the Jet Propulsion Laboratory in Pasadena, California. His work on the Voyager, Galileo and Lunar Polar Orbiter programs has earned him a number of NASA awards.

David F. Malin, Photographic Scientist at the Anglo-Australian Observatory in New South Wales, Australia, has pioneered techniques used in conjunction with optical telescopes for enhancing faint images from distant galaxies.

Dr. Stephen P. Maran is a Senior Staff Scientist at the laboratory for Astronomy and Solar Physics of the NASA-Goddard Space Flight Cen-ter. He is author of more than 200 professional articles and co-author of the volume *New Horizons in Astronomy.*

Dr. Harold Masursky is a Senior Scientist at the Branch of Astrogeology of the U.S. Geological Survey. He contributed to many NASA missions, among them Ranger, Lunar Orbiter, Surveyor, Apollo, Mariner, Viking, Pioneer, Voyager, Galileo and Venus Radar Mapper.

Dr. Laurence A. Soderblom is the Sherman Fairchild Scholar at the California Institute of Technology. A specialist in the geologic histories of the planets and moons of the solar system, he has participated in the *Mariner 9,* Voyager, Galileo and Venus Radar Mapper missions.

Dr. David J. Stevenson is a Professor of Planetary Studies at the California Institute of Technology. He received the Urey Prize, awarded by the American Astronomical Society in 1984 for his studies of the processes of planetary formation and the physics of planetary interiors.

Dr. Richard J. Terrile is on the technical staff of the Jet Propulsion Laboratory. He is involved in various programs of planetary astronomy and atmospheric and geological studies.

CONTENTS

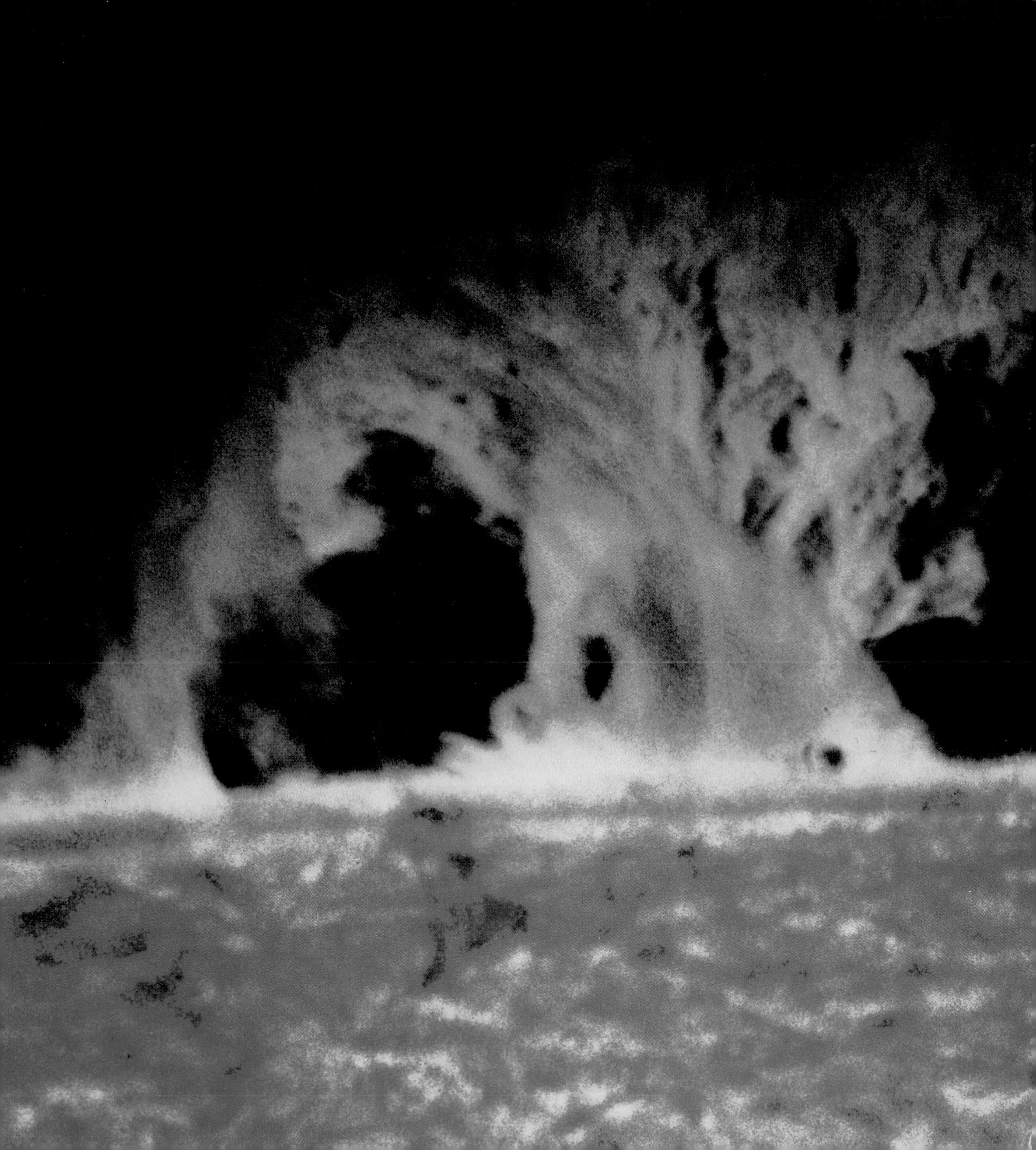

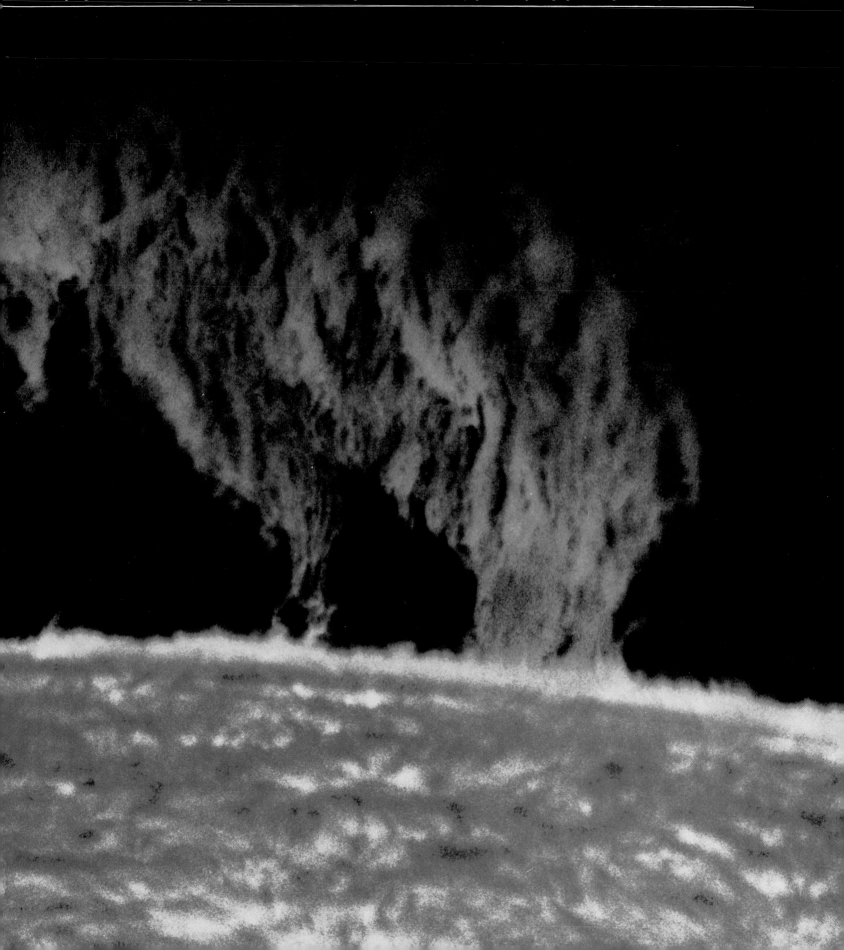

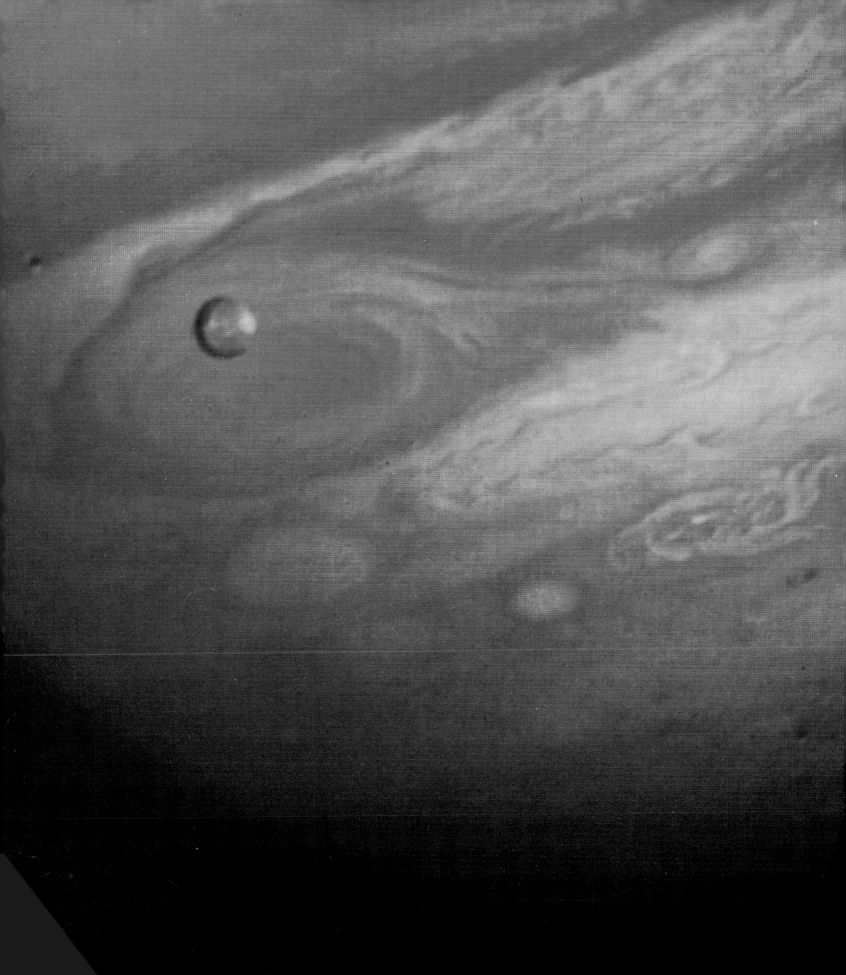

Wreathed in turbulent clouds, the giant planet Jupiter dwarfs two of its moons, Io *(left)* and Europa *(right center),* in this photograph taken by the U.S.

space probe *Voyager 1*. Below Io is Jupiter's Great Red Spot, an incessant storm so vast that it would blanket three Earths.

In this photograph taken by the *Voyager 1* spacecraft, the rings of Saturn — dense bands of orbiting particles — cast their shadows on the giant planet's

luminous cloud cover. At top, one of Saturn's 17 moons, Tethys, swings in its orbit 183,000 miles from the planet.

Comet West, a chunk of ice and cosmic dust several miles in diameter, speeds past Earth on March 9, 1976, trailing a cloud of ice crystals 30 million

miles long. Leaving Earth behind, the comet whirled past the sun and eventually left the solar system.

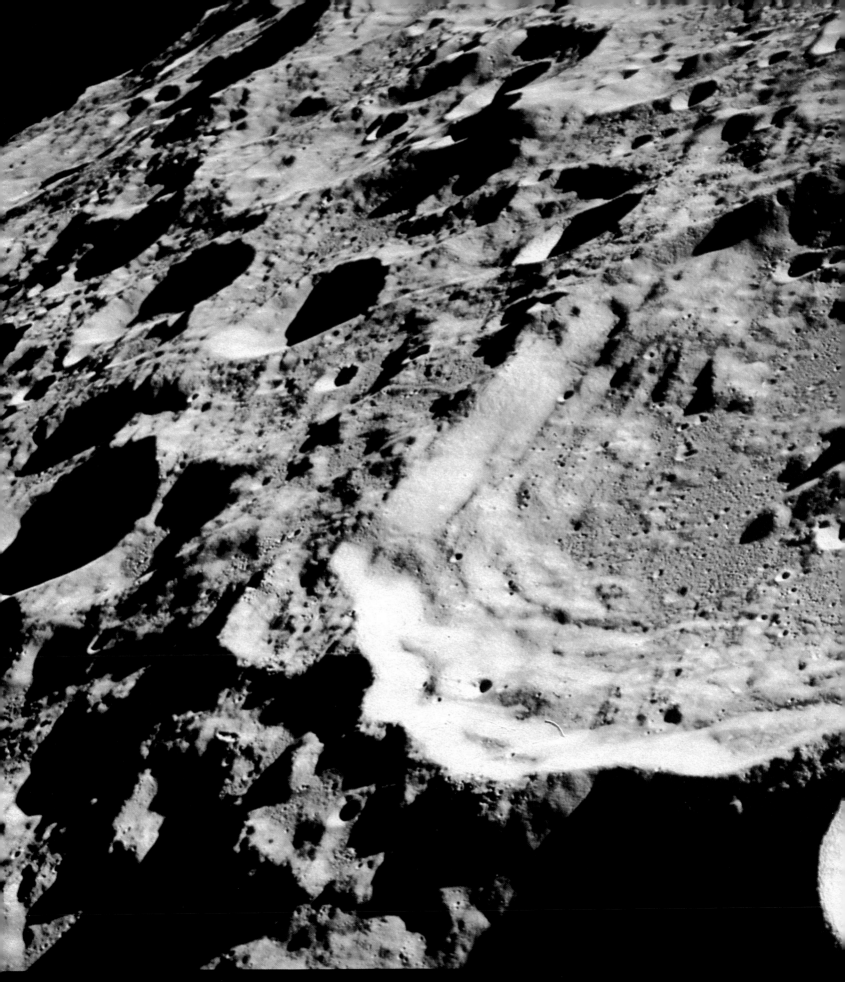

The far side of Earth's moon, photographed in 1969 from the *Apollo 11* spacecraft in lunar orbit, stretches away in a bleak vista of impact craters.

and fine dust. The large crater at center is about 30 miles across and, scientists believe, 3.5 billion years old.

THE SEARCH FOR DISTANT WORLDS

The thing hurtled down on Earth at 7:17 a.m., mystical and terrifying, to burst upon the forest people of Tunguska, in Siberia. No one knew what it was or where it had come from. They only knew the horror of the moment, the noise, the light and the awful heat.

A farmer named S. B. Semenov was sitting on the steps at the trading post that morning of June 30, 1908, when, by his word, "the sky was split in two, and high above the forest the whole northern part of the sky appeared to be covered with fire. My shirt was almost burned on my body. At that moment there was a bang in the sky and a mighty crash." The shock wave tossed Semenov from the steps and left him unconscious. Another witness near a river reported a blast and a hot wind "so strong that it carried off some of the soil from the surface of the ground, and then drove a wall of water up the river." Still another awe-struck farmer described the thing itself as "an elongated flaming object flying through the sky. It was many times bigger than the sun but much dimmer, so that it was possible to look at it with the naked eye. Behind the flames trailed what looked like dust. It was wreathed in little puffs, and blue streamers were left behind from the flames." This witness also heard great blasts, "and the ground could be felt to tremble, and the windowpanes in the cabin were shattered."

These people were lucky, for they were many miles from the heart of the cataclysm that erupted from the sky. Directly beneath the fireball, 800,000 acres of woodland ignited into an ocean of flame, the trees flattened so that their trunks pointed outward from the explosion like the spokes of an enormous wheel, while 500 grazing reindeer were roasted where they stood. The shock wave kept surging outward to knock down horses 400 miles away and ultimately to mark its violence on seismographic instruments around the world. The next few nights the skies in Western Europe glowed so brightly that Londoners reported being able to read their evening newspapers without artificial light.

Not until the age of the hydrogen bomb would mankind witness a detonation of comparable force and fire. Yet the Tunguska Event, as it came to be called, left no crater, nor any other obvious clue to what had happened — no shards of exotic metal from an alien spaceship, no odd stones or iron fragments to suggest that some random piece of sidereal matter had hit Earth.

Decades later, Soviet scientists reported the discovery at Tunguska of strange little diamonds of the sort sometimes found during laboratory analysis of meteors that had survived their impact with Earth. Many scientists

Studious astronomers, depicted in a 15th Century manuscript illustration, examine a stylized solar system based on the belief that Earth is at the center. In rotation above are the sun, the moon, and the planets Mars, Mercury, Venus, Jupiter and Saturn.

concluded that a comet—a kind of dirty empyrean snowball mixed with other space debris—had exploded from atmospheric friction, leaving behind nothing but particles like the diamonds. They calculated that the comet had probably been 100 yards across, that it had weighed about one million tons and that it had plummeted earthward at a speed of 70,000 miles per hour.

If it was a comet, or a fragment of one, then Earth had been visited not by an alien enemy but by a long-lost little brother, indeed a twin brother in time of birth. For many astronomers now believe that the sun, the planets and the moons of the solar system, together with the comets and all the other debris in this corner of the universe, came into being from the same matter and in the same cosmic era some 4.6 billion years ago.

The solar system began as a cloud, or nebula, of interstellar dust mixed with hydrogen and other gases floating out toward the edge of an ordinary-sized galaxy, the Milky Way, whose rotating swath of perhaps 100 billion stars makes up one of about 100 billion galaxies in the universe. Triggered by some external force, the nebula started to contract, developing gravitational pull and an increasingly dense core at the center. In reaction, the whole great cloud began rotating faster and faster, flattening the nebula and further increasing the density and the heat of the core. Eventually, after millions of years, the core ignited and became a natural nuclear furnace. Thus was born a new little yellow star approximately 864,000 miles in diameter—the sun. The nearest star in the Milky Way lay about 25 trillion miles away.

Farther out in the nebula, much smaller accretions of condensed gas and other cosmic particles formed into primordial planets, whose more modest density lacked the pressure to light a nuclear furnace. These bodies fell into orbits around the newborn star at distances averaging from 36 million to about 3.7 billion miles. Their paths were defined by a marvelous equilibrium between the sun's gravity and their own centrifugal force. Meanwhile, too, all of the bodies were rotating at various speeds on their axes—as the whole system moved majestically with the Milky Way on a regular round trip that would take about 230 million years.

The planets closest to the sun became predominantly solid and rocky as heat cooked off most of their internal gases, which were then blown into space by a wind of electrons and ions given off by the sun. Farther out, in the colder reaches, the planets and their moons were protected by distance; they stabilized as gigantic globes of liquefied gas, perhaps with solid cores and, in some cases, thick crusts of frozen substances. Along the edges of the new solar system floated frigid leftovers from the system's creation—dirty snowballs like the one that, by orbit or by chance, had headed earthward to explode above Tunguska.

On its journey, the comet probably took a long, elliptical tour of the solar system, passing, at different distances, the wondrous forms of the other members of the family. Among the more striking of these are:

● Saturn—a yellow-tan gaseous giant 75,000 miles in diameter, swept by winds of 1,000 miles per hour and festooned with an awesomely beautiful set of thousands of rings of ice and rock 620,000 miles wide but in places only 500 feet thick.

● Jupiter—a monster two and a half times as massive as all the other

Charred and flattened trees litter the remote Tunguska region of Siberia, devastated in 1908. The destruction was caused when a fiery body, believed to be a comet, exploded with the force of a 12-megaton nuclear blast.

planets combined, circled by its own system of 16 moons, with fantastic curtains of lightning crackling above its hydrogen atmosphere.

● Mars — a red ball of rock, its poles covered with ice, its arid landscape dotted with sand dunes, split by a canyon thousands of miles long and topped by a volcano 15 miles high and 250 miles wide.

● Earth — a pretty blue body nearly 8,000 miles through, temperate and moist, its crust rumpled and unstable, three fourths of the surface covered with water, the atmosphere composed mainly of nitrogen and oxygen — and sustaining the only known life in the solar system.

● Venus — a stifling place, where temperatures soar to 900° F. and a thick unbroken cover of pearl-colored clouds envelops the entire planet, dropping a perpetual rain of sulfuric acid.

● Mercury — a pockmarked cinder only 3,000 miles in diameter, its atmosphere burned completely away by the sun, rotating so slowly that 176 days pass from sunrise to sunrise, yet whirling so quickly around the sun that its year lasts only 88 days.

These are only a few of the many intriguing aspects of Earth and the planets nearest to it, the ones whose shining forms people have been able to pick out with the naked eye ever since the first intelligent human creature gazed up into the heavens. The rest of the planets lie well out of the range of natural sight, beyond the rocky belt of asteroids that orbits between Mars and Jupiter. These are the chill spheres of Uranus and Neptune and the little iceball Pluto.

Some of the solar system's fascinating features have only recently come to light, as science, with the aid of rocketry, got its feet off the ground to look at space from space for the first time. Others had turned up earlier, as optical and then radio telescopes were invented and refined. Yet there has never been a time when curious people did not try, by whatever means were available, to study the stars, watch the movements of the planets and — perhaps most of all — attempt to understand where the planet Earth fits in the grand scheme of the universe.

The first recorded conclusions about the solar system were set down by the Sumerians of the Middle East about 5,000 years ago. To them Earth was flat, motionless — and clearly the center of the universe. The heavens appeared as a tin dome within which the gods moved the stars, the sun and moon, and the five closest planets. Though the sizes of these glowing, supernatural bodies varied, they all seemed to move at about the same distance from Earth. Beyond the dome there was nothing. Other early civilizations — such as the Chinese, Babylonian and Egyptian — had much the same view of Earth as the hub of things. The moving heavens, including the faintly luminous band of the Milky Way, were the realm of gods and demons who caused the seasons to change, wars to be won and lost, and who otherwise influenced the currents of life on Earth.

Beginning sometime around 600 B.C., however, a succession of Greek philosopher-scientists began systematically to demystify the infinite puzzle of the sky. Aristotle, for one, concluded that Earth probably was round because its shadow was always round as it fell on the moon during a lunar eclipse. A brilliant but little-known natural philosopher named Aristarchus went much further — too far, in fact, for the mystical tastes of the times. Aristarchus, who lived from about 310 B.C. to 230 B.C., became a master of the mathematics called geometry, as well as a devout watcher of the skies. He noted the size of Earth's shadow on the moon during a lunar eclipse, then used clever reasoning combined with measurements of angles to deduce that the sun was much larger than Earth — and many times farther away than the moon.

Based perhaps on this certainty of the sun's predominant size, Aristarchus arrived at the radical conclusion that the sun — not Earth — was the center of the planetary system. In the course of his studies he went on to decide that the planets all orbited the sun, which he boldly catalogued as a star, and that the spherical Earth was simply one of the planets. What is more, he believed that Earth rotated on its axis.

Aristarchus set down these astonishing insights in various treatises, all but one of which have long since disappeared. The earliest surviving commentary on his work was written by the scientist Archimedes: "Aristarchus brought out a book consisting of certain hypotheses. His hypotheses are that the fixed stars and the sun remain unmoved, that the earth revolves about the sun in the circumference of a circle, the sun lying in the middle of the orbit."

Scarcely better known to posterity was another superb astronomer, named Hipparchus, who set up his own observatory on the Aegean island of Rhodes in about 150 B.C. There he constructed or assembled an arsenal of instruments for measuring the relationships between celestial bodies and the sizes of the nearest ones. His instruments — among them astrolabes and armillary spheres — permitted him to chart the positions and relative brightness of 850 stars; calculate the distance between Earth and the moon to within a few miles of the actual figure (238,857); and work out a system for predicting the positions of the sun and moon the year round.

Yet neither of these men won the enduring reputation and credence that went to the last of the great ancient philosopher-scientists — who, unfortunately, managed to be wrong. He was Claudius Ptolemaeus, better known simply as Ptolemy, who flourished in the Second Century A.D. Ptolemy early fell in love with the study of the heavens and tended to express his passion in fulgent prose. "Mortal as I am, I know that I am born for a day," he wrote. "But when I follow at my pleasure the serried multitude of the stars in their circular course, my feet no longer touch the earth."

Nor did they. Much of what Ptolemy and his contemporaries espoused about the skies was pure astrology, a codification of Egyptian and Babylonian beliefs that the aspects of the planets and stars controlled earthly matters down to the color of a person's skin and even the outcome of a romance. Rejecting the ideas of Aristarchus, Ptolemy insisted that a stationary Earth stood at the heart of the universe, with everything else moving about it. After all, went Ptolemy's logic, if Earth spun around, birds could not hold on to their perches.

Nevertheless, Ptolemy was a tireless observer. He named the stars and the constellations, and recorded endless sightings. Around his central Earth he constructed a model universe in which all the bodies moved in progressively larger concentric circles. Like other men of his time, he knew that while the stars traveled across the night sky in identical paths and patterns, the planets did not. Sometimes the planets even seemed to wander backward, or retrogress, in the heavens, and Mars performed a periodic loop-the-loop. Ptolemy accommodated these peculiarities by embellishing the planets' circular orbits with epicycles — little suborbits that accounted for the irregular movements.

Framed by the massive pillars of Stonehenge, the sun rises above a pointed rock on the summer solstice. From such evidence, many astronomers have concluded that Stonehenge, completed more than 4,000 years ago, served as a celestial observatory, keeping track of the seasons through the movement of the sun.

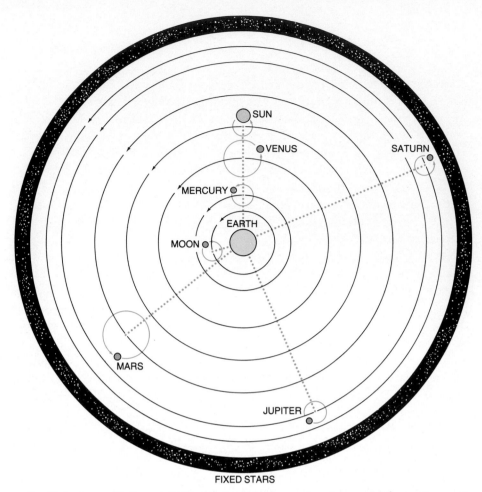

The Second Century A.D. scholar Ptolemy *(above)* formalized the long-held view that the sun, moon and planets traveled in perfect circles around Earth *(left)* and inside a sphere of fixed stars. To explain the planets' noticeable deviation from the theoretic orbits, Ptolemy suggested that they were moving in small circles, called epicycles, at the same time that they revolved around Earth.

It all fit beautifully with his observations, so that he could foretell exactly where the planets would be at any time. He recorded his concept of the cosmos in 13 volumes that came to be called the Almagest, or "the Greatest." This work, which also summed up the astronomical achievements of the past, survived the collapse of Greece and the destruction of the centers of Greek culture. It continued to endure after the fall of Rome and throughout Europe's Dark Ages to become the backbone of the Roman Catholic Church's dogma on the nature of the universe. That is, by God's design Earth lay at the center of things, unmoving, and the heavens as defined by Ptolemy were perfection, complete.

Few dared to say otherwise, even during the intellectual ferment of the Renaissance, for fear of imprisonment, torture, death or all three at the hands of papal inquisitors. The challenge finally came from, of all people, a Roman Catholic canon named Mikolaj Kopernik, who lived in a tower adjoining the cathedral at Frombork, Poland. Kopernik had been born into a prosperous merchant family in 1473, a time when learning and culture were flourishing in Poland. He studied first at the celebrated University of Cracow, of which a contemporary scholar wrote, "All sorts of proficiencies are practiced, such as the study of speaking, poetics, philosophy and physics. But the science of astronomy stands highest there."

In 1496, Kopernik went to Italy and spent a decade studying canon law, medicine and the stars. Returning home to serve as secretary and physician to his uncle, the bishop, he developed some radical notions about the order of the heavens. "Although the idea seemed absurd," he noted, "I began to think of a motion of the earth." The idea would not have seemed in the least absurd to certain Greeks. Neither would a great deal else of what Kopernik began to record in notes and private manuscripts. In 1512

Copernicus, who challenged the traditional view of the solar system in the 16th Century, moved the sun to the center and placed Earth — with its companion moon — among the revolving planets *(right)*. Although he was correct in this essential respect, Copernicus did not realize that the orbits of the planets were elliptical; he retained Ptolemy's fanciful epicycles and perfect-circle orbits.

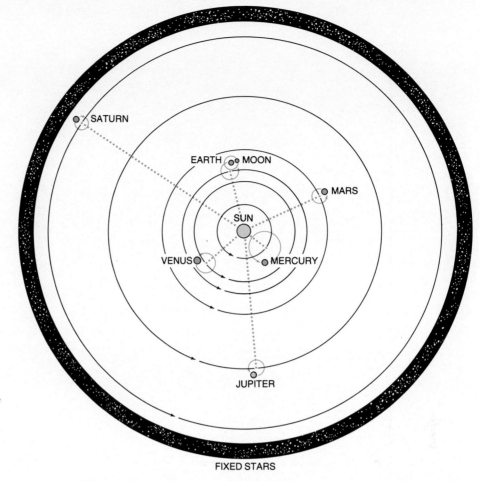

he retreated with his growing pile of material to the Frombork Tower, which he affectionately referred to as *in remotissimo angulo terrae* — in the remotest corner of the earth.

For 20 years Kopernik studied the heavens and compiled his observations. Several times along the way he was on the verge of giving up. "The scorn which I had to fear on account of the newness and absurdity of my opinion," he wrote, "almost drove me to abandon a work already undertaken." Nevertheless, he struggled on to complete the book in 1533, then hid the manuscript in his house, sharing it hesitantly with only a few friends during the next 10 years.

The book, which he signed with the Latin version of his name, Copernicus, bore the title *De Revolutionibus Orbium Coelestium — On the Revolutions of the Heavenly Spheres.* It began, very diplomatically, with a dedication to Pope Paul III. It made humble references to the Most Holy Father and to "the godlike circular movements of the world and the course of the stars."

Having established the piety of his motives, Copernicus got on with the message. "As if seated upon a royal throne," he wrote, "the sun rules the family of the planets as they circle around him." Copernicus then put the godlike Earth back among the rest of the planets, where Aristarchus had positioned it 1,800 years before, and confirmed that the only thing revolving around Earth was the moon. Furthermore, Earth rotated on its axis, causing, among other phenomena, the alternate conditions of day and night.

Up to this point *De Revolutionibus,* though filled with ideas that no one else at the time dared state, really was Aristarchus revived. In fact, the original manuscript of *De Revolutionibus* bore a note in Copernicus' neat

longhand crediting Aristarchus with this shocking concept of the solar system. Copernicus went on to position the planets in their proper places and get them moving at the right relative speeds within the system. He correctly placed Mercury closest to the sun and circling it in a quick little orbit; next came Venus, then Earth, Mars, Jupiter in progressively slower orbits, and finally Saturn, performing a stately circle of many years.

These aspects of the solar system, together with some elegantly documented arguments in support of them, helped to account — at least in theory — for some of the wandering tendencies of the planets. For example, the outer planets seemed to move in retrogression because Earth, in its smaller orbit, revolved more rapidly than they around the sun, so that they appeared to travel backward against the far-distant background of the stars.

Copernicus did not publish *De Revolutionibus* until 1543, by which time he was 70 years old and much enfeebled by a stroke. Some accounts say that when the first copy reached him, he was bedridden and lay only moments from death. Indeed he may never have seen the final work — from which, for reasons never explained, the credit to Aristarchus had been deleted. In any case, Copernicus did not live long enough to suffer in the centuries-long battle that his work triggered between church and science, a battle that was to cost science some notable casualties. Throughout the struggle, and forever afterward, *De Revolutionibus* would stand as the lodestar among all works on astronomy. Although neither entirely original nor entirely correct, it comprised the best documentation yet done on the solar system.

The major flaw in *De Revolutionibus* was the retained belief that the planets orbited in circles. They do not, and Copernicus had to include some epicycles in his orbits to account for the more peculiar wanderings by the planets. But the mistake barely survived the century. For by the year 1600, astronomy verged on a period of discovery and invention that, for its time, was almost as dynamic and revelatory as the modern space age. This despite the fact that, by 1600, the Church had become enraged at scientific meddling with its dogma, to the point of burning alive a scholarly Catholic cleric named Giordano Bruno for daring to suggest that the universe might contain other inhabited worlds.

Within a few years of Bruno's immolation, the Church had a lot more radical ideas to worry about. Some of them came from a brilliant German Lutheran schoolteacher and astronomical theorist with a classroom demeanor so vague and boring that his pupils either fell asleep or walked out. Were it not for a stroke of fortune, this professor, Johannes Kepler, might have vanished into the mists of history. Luckily, however, he came into possession of some priceless data on planetary movements developed in Prague by a fast-living Danish stargazer named Tycho Brahe, who wore a false nose of gold and silver to replace the real one lost in a duel.

Kepler had constructed a brand-new model of his own for the solar system, based on a bewildering theory of the relationship between the planetary orbits and five geometric forms. He sent Brahe a copy of his theory of circular orbits, expecting the noted Dane's data to confirm it. At first Brahe declined to share his data, but he invited Kepler to come to Prague and work as his assistant. In 1600, Kepler accepted.

Within a year, Brahe had eaten and drunk himself into an early grave,

Pointing skyward in his lavish observatory near Copenhagen, the 16th Century Danish astronomer Tycho Brahe directs a pair of assistants as they record his sightings. While rejecting the notion of a sun-centered solar system, Brahe conceded that all the planets except Earth revolve around the sun, and he made many useful measurements of their orbits.

and Kepler fell heir to the Dane's lifetime accumulation of meticulous observation. He discovered that Brahe's data, especially that relating to the troublesome movements of Mars, indicated that the correct orbital shape for planets was an ellipse. Kepler said in dismay, "What a foolish bird I have been." But he embraced the concept and went beyond.

Eight years later, in his book *Astronomia Nova (New Astronomy),* Kepler published two laws he had based on Brahe's observations of the orbit of Mars. The first law established that every planet moves in an elliptical orbit with the sun as one of the foci, the other focus being a theoretical point in space. The second stated that all the planets speed up in their orbits as they approach the sun and slow down proportionally as they get farther away.

Later Kepler propounded a third law, in a book titled *The Harmonies of the World,* showing a direct mathematical relationship between the cube of a planet's distance from the sun and the square of its orbital period. By applying this formula, Kepler was able to calculate the relative distances between the sun and the planets even though the actual distances had not yet been worked out accurately. He took the mean distance between Earth and the sun as his basic yardstick — one astronomical unit. Then, using the known orbital periods of Earth and Mars, 365 days and 687 days respectively, he calculated correctly that Mars was 1.523 astronomical units from the sun. The actual Earth-to-sun distance would remain uncertain until the 20th Century, when radar-wave measurements set the figure at 93 million miles.

These three principles correctly explained most of the discrepancies in the planets' movements. They also removed the Lord's perfect circles from the heavens and suggested that perhaps something besides God's hand might be responsible for the way things behaved up there. "My aim is to show that the celestial machine is to be likened not to a divine organism," Kepler wrote, "insofar as nearly all the manifold movements are carried out by means of a single, quite simple magnetic force." This was powerful stuff. Not only was Kepler defying his church's doctrine, but by the reference to magnetism he was inching, intuitively yet unknowingly, very close to another universal law, which one day would be called gravity. "Astronomy," he said, "is part of physics." For his heresy Kepler was excommunicated, and he died penniless in 1630.

An Italian contemporary named Galileo Galilei, who also challenged the cherished Ptolemaic creed, suffered even more for that offense. The year Kepler's *Astronomia Nova* came out, Galileo was teaching astronomy and mathematics at the university in Padua, where he lived with his longtime mistress and their three children. That same year Galileo heard of the Dutch invention of the spyglass, a succession of lenses arranged in a tube so that distant objects were made to appear much closer. "This seemed to me so marvelous an effect," wrote Galileo, "that it gave me occasion for thought." He immediately began building his own spyglasses, the best of which magnified up to 32 times.

Galileo's splendid new instruments produced staggering results. The astronomer saw that Venus, like the moon, went through a full range of phases from crescent sliver to practically full. In Ptolemy's complex system of epicycles, with the sun always outside Venus as they both revolved around Earth, Venus could never wax beyond a narrow crescent. Now, by firsthand visual evidence, Ptolemy was proved wrong.

Basing much of his work on observations made by his colleague Tycho Brahe, the German mathematician Johannes Kepler determined that the planets swing in elliptical orbits around the sun. He also demonstrated that the orbiting bodies speed up as they near the sun and slow down as they move away. The effect of the sun on the planets' orbital speed explains why Mars *(below)* travels from point A to point B in the same time that it takes to cover the much shorter span between C and D.

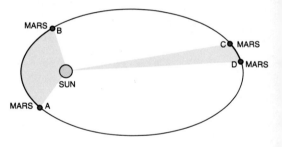

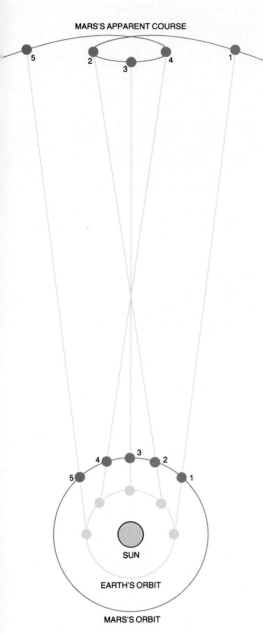

MARS'S APPARENT COURSE

EARTH'S ORBIT

MARS'S ORBIT

In showing that the planets change their speeds in orbit, Johannes Kepler demonstrated that the epicycles posited by Ptolemy were illusions. In the diagram above, the actual positions of Mars and Earth are shown at the bottom; at the top is Mars's apparent course, as viewed from Earth. Because its orbit is closer to the sun, Earth moves more rapidly than does Mars; thus, Mars seems to move backward at positions 3 and 4, and then to move forward once more when it reaches position 5.

The revelations mounted during the week of January 7, 1610, while Galileo was peering at Jupiter through his glass. "I noticed three little stars, small but very bright, were near the planet," he wrote. Over the ensuing nights Galileo watched in fascination as the newfound bodies rotated around Jupiter. And that was not all: "These observations also established that there are not only three but four erratic sidereal bodies performing their revolutions around Jupiter like the Moon about the Earth, while the whole system travels over a mighty orbit about the Sun in the space of twelve years."

As Galileo pursued his astronomical research, his telescope revealed that the sublime sun had spots on its face — and they moved, indicating that the sun rotated. The moon, supposedly smooth, was pocked with craters and mountains. He saw Saturn's rings but could not figure out what they were. He apparently sighted Neptune several times but did not realize that it was a planet.

Although these stunning discoveries obviously ran counter to Roman Catholic doctrine, Galileo seemed unconcerned at first. "The Bible shows the way to go to heaven," he commented in a letter to one of his disciples, "not the way the heavens go." The letter eventually ended up in the hands of the Inquisition. When Galileo heard that he was being investigated, he decided to go to Rome to clarify his position. Nevertheless, in February 1616, the Holy Office in the Vatican declared the sun-centered theory to be "foolish and philosophically absurd" and also banned sales of Copernicus' *De Revolutionibus,* which Galileo had openly admired. Galileo was ordered to stop teaching heresies. The astronomer said that he would comply — but he kept on compiling his thoughts in a book that he attempted to publish some years later, in 1630.

The Church censors demanded revisions. Galileo, now in his late sixties and physically infirm, cried out in despair. "My life wastes away," he mourned, "and my work is condemned to rot." But after tinkering with the preface in an attempt to mollify the censors, he published the book in 1632 under the title *Dialogue on the Two Chief World Systems.* Pope Urban VIII was furious, and he ordered Galileo to Rome to face a full-dress inquisition, despite the protests of a sympathetic official who declared Galileo to be so sick that he might die en route. In Rome, the aged scientist was commanded to recant the whole concept of a heliocentric planetary system. The sick old man caved in. "I do not hold and have not held this opinion since the command was given me that I must abandon it," he quavered.

Satisfied but totally unsympathetic, the Pope sentenced Galileo to a lifetime of house arrest at the astronomer's home in a village outside Florence. There Galileo lived and worked on for nine years more, growing blind from his illness and bitterly denouncing his judges for their "hatred, impiety, fraud and deceit." He died in 1642, and more than a century passed before Rome brought itself to accept his visible proofs — or Kepler's physical laws — on the nature of the solar system.

By then, more powerful proofs had been set forth in the English country town of Woolsthorpe. The proofs came from a man of towering intellect who had created integral and differential calculus by the time he was 23. Isaac Newton spoke little but observed a great deal. He studied the sun until his eyesight failed, and then had to sequester himself in a darkened room for three days before he regained his sight. He watched the moon circling Earth. He watched thrown objects sink to the ground as they lost their speed. And legend has it that he watched with particular care an apple that fell from a tree in his garden; he realized that the same force that made the apple fall kept the moon moving in its orbit. From all his observations and much, much more, Newton deduced certain fundamental laws that govern the behavior not only of the bodies in the solar system but of the entire universe.

He revealed the laws, together with their mathematical proofs, in a book entitled *Philosophiae Naturalis Principia Mathematica,* printed in 1687 with Newton's modest announcement: "I now demonstrate the frame of the system of the World." And he proceeded to do just that. Most important, he demonstrated that objects, such as the sun and the planets, or Earth and an apple, attract one another in direct proportion to their size and in inverse proportion to the distance between them. He called this attractive force gravity. He also established that an object in motion would keep moving at the same speed and in the same direction unless some force — such as gravity — diverted it.

Thus the speed and mass of the moon are counteracted by the force of gravity between the moon and Earth, creating a perfect equilibrium that holds the moon in its orbit. The same is true of all other orbiting bodies in the skies. Asked how he had managed to deduce the law of gravity, Newton replied, "I gathered it from Kepler's theorem about 20 years ago." And to the question of how he arrived at his other brilliant insights, he said simply, "By thinking upon them." By such elemental processes did Isaac Newton explain the workings of the solar system, unite the universe with laws that apply in the heavens as well as on Earth, and change the course of scientific thought forever.

On his death in 1727, Newton also left as a legacy to science a new reflecting telescope, with a concave lens and a flat mirror, that provided sharper images than did the lenses of previous instruments. Using Newton's design, a Prussian émigré to England, William Herschel, developed a formidable instrument with a 7-foot focal length and a 6.2-inch aperture; he showed it to astronomers at Greenwich Observatory and proudly reported that they "declared it to exceed in distinction and magnifying power all that they had seen before." And on March 13, 1781, he launched the modern era of planetary discovery.

That evening Herschel was scanning a field of stars when, between 10 p.m. and 11 p.m., he spied something odd. "I perceived one that appeared visibly larger than the rest," he recounted. He knew the stars were so far away that, even with the best telescope of his day, they never appeared larger than mere points of light. But when he changed the eyepieces, doubling the magnifying power and then doubling it again, he saw that the mysterious body could not be a star.

At first Herschel thought the new object to be a comet. But after lengthy study, his associates argued that it was a planet. He conceded that its color was "not fiery or cloudy like that of comets but of an equable bright uniform lustre, something between the light of Jupiter and Venus." Moreover, its slow speed and direction were consistent with that of the planets. Subsequent measurements showed that the body was moving in an orbital path about twice as far from the sun as Saturn's.

Herschel had found a new planet, the first one discovered since ancient

Ingenious Tools for Pioneer Astronomers

In the years before the telescope was invented, astronomers developed a number of instruments to record and predict the movements of heavenly bodies. Most of these devices, among them the astrolabes, quadrants, nocturnals and sundials shown here and on the following pages, were used to make celestial measurements; they operated on mathematical principles and were generally known as mathematical instruments. Tabletop planetariums were constructed as working models of the solar system; astronomers placed the planets in their relative positions in order to study their orbits and their interrelations.

The pioneering tools were both accurate and beautiful. For example, the exquisite astrolabe pictured at right, fashioned in 1556, is also a sophisticated computer. Using it, a scientist could determine the correct time of day or night and project the rising and setting time of the sun for any day of the year.

This 16th Century brass astrolabe was employed, in complicated sequence, for a variety of astronomical calculations. An astronomer could use it to find out when a star rose and set even though those events occurred during the day, when the star was invisible. First, at night, he sighted along the transverse arm of the astrolabe, measuring the star's angle of altitude. He rotated the openwork plate until the pointer representing that star was aligned with the corresponding altitude marked on the instrument's outer edge. This set the prevailing pattern of the star's movement. Finally, a calendar scale on the back would show the astronomer when the target star rose and set.

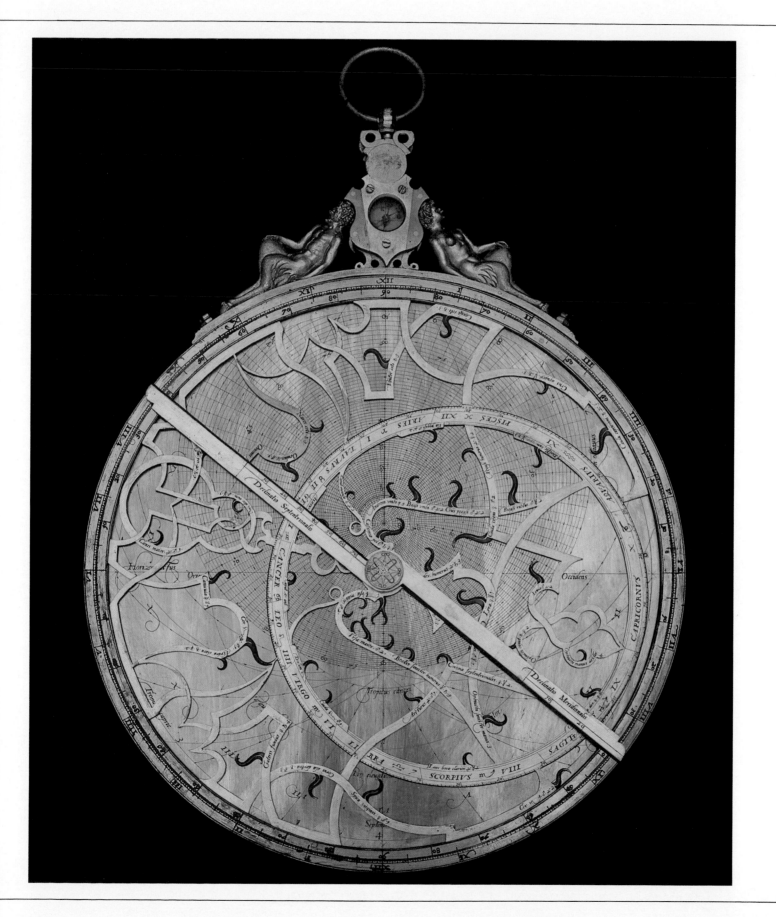

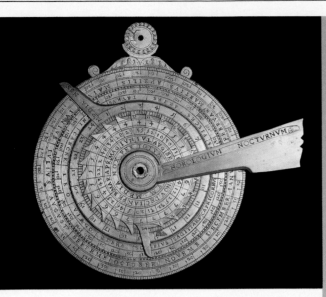

This 16th Century nocturnal tells the time at night, basing its reading on the relative positions of three major stars.

Earth is at the center of this 16th Century armillary sphere *(right)*, a model of the presumed orbits of the sun and planets.

Made of gilded brass, a portable German sundial is equipped with a compass to ensure correct alignment with the sun.

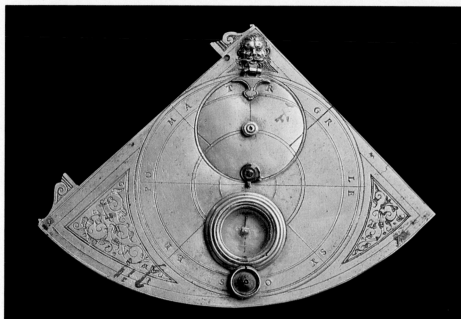

This gilded quadrant, fitted with a small compass and a sundial, was once the property of Pope Gregory XIII, whose interest in astronomy led him to urge a revised calendar in 1582. Quadrants were so called because of their quarter-circle shape and were used to navigate and to tell time by the stars.

Measuring more than three feet across, a planetarium made in England in the 18th Century depicts the solar system as it was known at the time. A clockwork mechanism moves the planets in orbits around the central sun.

times. He wanted to name the new planet Georgium Sidus (the Georgian Star) after his new patron, King George III of England. But other astronomers insisted on sticking to mythological roots, and eventually they named it Uranus, for Saturn's father.

Before long, Herschel had more news to relate about the skies. After studying the changing positions of the stars, Herschel announced that the sun did not sit at a fixed point but traveled through space, carrying the planets along with it. Moreover, with his marvelous telescopes he defined the starry make-up and approximate shape of the Milky Way. And pairs of distant stars rotating about each other confirmed Newton's axiom that gravity enforced its power everywhere in the universe. All these tidings, along with the codification of Newton's laws, would reverberate indefinitely throughout the world of science. Together with advancing technology and sophisticated new mathematical techniques, they opened up the way to spectacular celestial discovery.

In fact, the next great breakthrough in planetary discovery had more to do with mathematics than telescopy. The groundwork was laid when astronomers, anxious to understand more about Uranus, began to calculate its orbital position backward in time. Studying old records, they were fascinated to find several instances when Uranus actually had been seen and recorded prior to Herschel without the observer's realizing that the object was not a star. Even more intriguing was the fact that several of these earlier observations, recorded with meticulous care, showed the planet slightly off its mathematically correct orbital path.

If the old observations were as accurate as they seemed, then some force might well be perturbing Uranus, bumping it out of its proper orbit. Several astronomers suggested that the force could be gravity from yet another unknown planet. And by the end of the 1830s they felt that the data they had pulled together might enable them to calculate the approximate position of this phantom planet, which would make a telescopic search for it much easier. Two men, each unaware of the other's efforts, set themselves to that task. One was a graduate student at Cambridge named John Couch Adams, the other a mathematical astronomer in France, Urbain Jean Joseph Leverrier.

In September of 1845 Adams had pinned down the phantom's theoretical position and took his data to Greenwich's ruling Astronomer Royal to request a program of observation. Unfortunately, the A.R., as he was known, was away from home twice when Adams arrived, and he was at dinner on Adams' third visit. In a daffy scene, the A.R.'s butler forbade this mere graduate student from interrupting the great man's repast. Adams retreated, leaving a copy of his calculations. But when the A.R. requested more information, young Adams considered the query irrelevant and chose not to respond. A year later, Adams did reply, and the A.R. eventually urged a search for the unseen planet.

Meanwhile in France, Leverrier found himself bogged down in the same kind of farcical situation. Though he, too, had his data in good order, he could not persuade any ranking observatory director to take him seriously enough for a concerted telescopic search. In frustration Leverrier wrote to a colleague, Johann Gottfried Galle, at the Berlin Observatory. Galle received the letter on September 23, 1846, and that very night persuaded his observatory director to let him use the telescope.

The great Italian astronomer Galileo built these two telescopes in the early 17th Century. The instrument on the right is 4½ feet long and magnifies 14 times; the shorter one, having a more powerful lens in the eyepiece, magnifies as much as 20 times.

Using his telescopes to study the planet Venus, Galileo finally overturned the notion of an Earth-centered solar system. According to the traditional, pretelescope view *(below, left)*, the sun always moved in orbit outside Venus and closely paralleling the planet; the observer, staring almost into the sun, could see Venus only at dawn and dusk, and only as a narrow crescent. But Galileo, with his telescopes, saw that Venus actually goes through a series of moonlike phases *(below, right)*, including a full face. This proved that Venus passes in orbit behind the sun and thus established that the sun was the center of the solar system.

When the instrument was focused on the position in the sky Leverrier had calculated, Galle at first saw nothing — but then suddenly he spotted a bright object that did not fit existing star charts. "That star is not on the map," exclaimed a volunteer assistant who was helping Galle. They followed it until it set, then tracked it all the next night, certain now that they had found something new.

And so they had. The bright object was the phantom planet, and it had been found within one degree of the point predicted by Leverrier. "The planet whose position you have pointed out actually exists," Galle wrote excitedly to Leverrier on September 25. Leverrier, who generally received credit for the discovery, lobbied to have the planet named after himself. But it, too, received a name from mythology — Neptune. Once again the solar system had been expanded, this time by more than a billion miles; the distance that Neptune lay beyond the orbit of Uranus.

As before, the discovery of another planet helped fuel the search for more revelations in space. Among the most diligent seekers was Percival Lowell, an elegant Boston Brahmin and a considerable success as a businessman, diplomat and writer. Lowell in 1877 was electrified to read that an Italian astronomer, Giovanni Schiaparelli, had reported seeing *canali* on Mars. The Italian word means "furrows," or "channels." But it came across in English to Lowell as "canals" — which, of course, implied construction by beings of advanced intelligence. Lowell was enthralled with the possibility that there might be life beyond Earth. And when he learned in 1893 that Schiaparelli was abandoning work because of failing eyesight, Lowell dedicated the remainder of his days to a passionate study of the skies, with two objectives: to find evidence of life on Mars, and to seek out yet another new heavenly body, which he referred to as Planet X.

With his own money he built a complete modern observatory outside Flagstaff, Arizona, on a prominence named Mars Hill. For more than 20

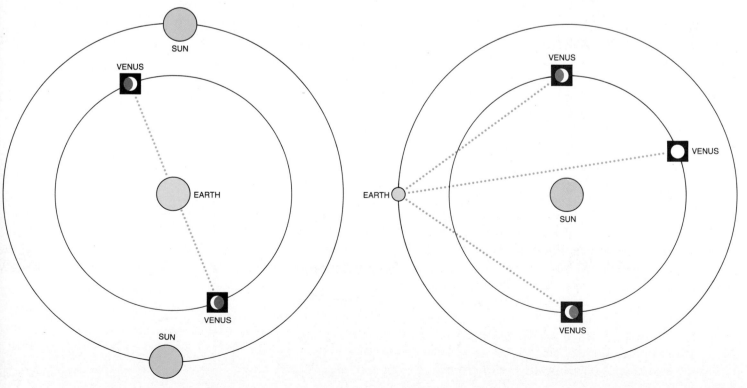

years he peered through the observatory's 24-inch telescope and found wishful confirmation of what he believed to be there. He wrote that Mars had a "reticulated canal system embracing the whole planet," and that this network, which distributed the planet's meager water supply from the polar icecaps, proved "the world-wide sagacity of its builders." It was all fascinating, but so devoid of scientific evidence that some of his contemporaries began to snicker.

Lowell's labors on behalf of Planet X were almost as exhaustive and, if anything, even less satisfying. He tried persistently to chart the orbit of the putative planet by studying the effect it seemed to have on the paths of Neptune, Uranus and certain comets. Then he used his calculations to help guide further observation and photographic study. Though the results were discouraging, he kept searching right up to his death in 1916, whereupon the Lowell Observatory temporarily turned its attention elsewhere in the skies.

However, several of Lowell's colleagues also suspected that Planet X really might be out there. And in 1929 the Lowell Observatory again took up the search. This time the zealot was Clyde Tombaugh, a 22-year-old one-time farmer with no advanced education. As a boy in Kansas, Tombaugh had built a 400-power telescope from a description published by *Scientific American* magazine. He drew pictures of what he saw through it of Jupiter and, on an impulse, sent them to the Lowell Observatory. The man who had become director, Dr. Vesto Melvin Slipher, was impressed and invited Tombaugh to come try out as an assistant. "I couldn't fail," Tombaugh later recalled. "I only had enough money for one-way train fare."

Hired that April at $90 a month, he labored for as long as 15 hours a day at what most people would have considered the world's most tedious job. In the renewed hunt for Planet X, the observatory had started taking thousands of telephotographs of small segments of the sky; pictures of each section were exposed several days apart. Each matched pair was then placed in a stereoscopic device called a blink comparator, which caused one plate and then the other to be illuminated in rapid succession. Since the plates exactly coincided in the viewfinder, all the fixed stars remained in the same positions with each blink. But a planet would have moved in the time between the two original exposures, and the blinking device would reveal its motion against the steady background of stars.

Each plate had tens of thousands of star images. Some had more than a million. The operator would examine one small region of the twin plates at a time, blinking them on and off, on and off, searching for the supposed planet. "You could easily go nuts," said Tombaugh. "No professional astronomer would go through that, so they wanted an enthusiastic amateur. With me, they got one."

Tombaugh divided the plates into thin horizontal strips and moved slowly across each strip, methodically blinking the star images. He toiled into the fall and through the winter with both the blinking and the photography. "I'd damn near frozen my fingers on the telescope," he recalled. "You couldn't have any heat or it would disturb the pictures. You are hunched over the telescope till 3 a.m." Then he would take his plates to the blink comparator and flash them, as he described it, "strip by strip, panel by panel. I was a perfectionist."

On the morning of February 18, 1930, Tombaugh began blinking a pair

In 1687, the brilliant Isaac Newton proposed his laws of motion, which explain the forces that keep orbiting bodies traveling in their regular courses. As shown in the diagram below, the moon's orbital momentum (*red arrows*) is counteracted by Earth's gravitational pull (*black arrows*), maintaining an equilibrium that holds the satellite in its orbit.

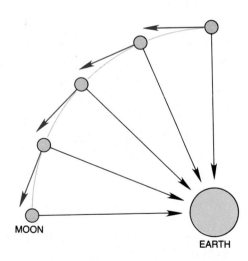

MOON

EARTH

Swathed in white clouds, Earth rises over the lunar horizon in this dramatic view from *Apollo 11*, in orbit around the moon. The moon orbits Earth at the rate of 50,800 miles per day, even as Earth covers 1.5 million miles a day in its travels around the sun.

of plates taken six days apart, and by afternoon he was about one fourth done. He turned the next small field of the plate into view. Suddenly, he spied an image that moved about three millimeters in the rapidly alternating views. "That's it," he exclaimed to himself.

"The object was far beyond the orbit of Neptune, perhaps a thousand million miles beyond," Tombaugh said. "A terrific thrill came over me." He rechecked everything, and triple-checked. Finally he walked down the hall to the office of director Slipher. "Trying to control myself, I stepped into his office as nonchalantly as possible. He looked up from his desk work. 'Dr. Slipher, I have found your Planet X.'"

That night was too cloudy for more picture taking. So to celebrate his achievement, the only person to discover a planet in the 20th Century walked down the hill to Flagstaff and saw a movie — *The Virginian*, starring Gary Cooper.

The new planet was named Pluto, for the god of the underworld. Orbiting at an average distance of about 3.7 billion miles from the sun, Pluto was the most remote planet known. However, some scientists believe that yet another planet, a 10th one, is lurking in deep space beyond the orbit of Pluto. For Pluto has turned out to be just a tiny speck of water, ammonia and methane ice with only about $\frac{1}{400}$ the mass of Earth. A few astronomers regard it as no more than an overgrown comet, far too small to exert any significant gravitational tug on Uranus or Neptune, whose masses are 15 and 17 times greater than Earth's, respectively. Nevertheless both of these giants have displayed unexplained deviations in their orbits. The path of Neptune in the 1980s, for example, deviates significantly from the orbit calculated for it by the U.S. Naval Observatory Nautical Almanac Office just 10 years earlier.

Many astronomers doubt that these aberrations are caused by another massive, undiscovered planet. Clyde Tombaugh, with his usual tenacity, kept looking for another planet for 13 more years after his discovery of Pluto. That search covered 70 per cent of the sky. "I would guarantee that within the area I examined there is no 10th planet," he said. But if there is another planet in the remaining portion of the sky, its chances of being found have improved exponentially in the decades since Tombaugh first began tinkering with his homemade telescope back in Kansas.

For in that eyeblink of cosmic time, scientists have developed infinitely more sophisticated devices for studying the skies, and they have learned more about what is out there than in all the preceding millennia. Telescopes quickly grew to 500-ton monsters with 100- and 200-inch lenses. Through these enormous eyes astronomers learned that various vague spiral nebulae, long thought to be mere gas clouds within the Milky Way, actually were far-distant galaxies. With this revelation, the boundaries of the universe came tumbling down, as those of the solar system had done.

Other scientists in the late 1930s discovered that a new instrument called the radio telescope could pinpoint the source of radio waves emitted by unseen stars and could even analyze some aspects of the stars' behavior and make-up. Laboratory experiments using electronic timers proved that light, in a vacuum, travels 186,000 miles a second, or six trillion miles in a year. With the speed of light as a highly accurate yardstick, scientists could measure the immense distances of interstellar space by timing the journey of radio waves bounced off nearby celestial bodies. By such means, both opti-

Anatomy of the Solar System

The solar system is an orderly community of nine planets, 44 moons and myriads of asteroids, comets and other small bodies, many of them sweeping in regular orbits around the central sun. The nature, properties and relationship of the bodies are anatomized in the data chart at right and in the orbital diagrams on the following pages.

As stars go, the sun is not particularly large. But in comparison to the planets it is gargantuan, as suggested by the sketches at right. The diameter of the sun, about 864,000 miles, is nine times Jupiter's, and in turn, Jupiter and Saturn dwarf the other planets, having diameters more than 11 times and nine times that of Earth.

The distances between the sun and the planets are simply too great to be shown here in scale. If the 3.7 billion miles between the sun and Pluto were reduced to one mile, the size of the sun would be 14.5 inches across and Pluto would be represented by a barely visible speck. In other terms, if Pluto were the size of a quarter, the sun would have to be 38 feet wide and placed about 31 miles away.

This chart lists the vital statistics of the planets. In three closely related categories, mass refers to the weight of a planet in relation to Earth's; density is a measure of the planet's mass as related to its volume; and volume expresses the quantity of three-dimensional space contained in the planet as compared with that in Earth. The greater a planet's mass, the stronger is its gravitational pull.

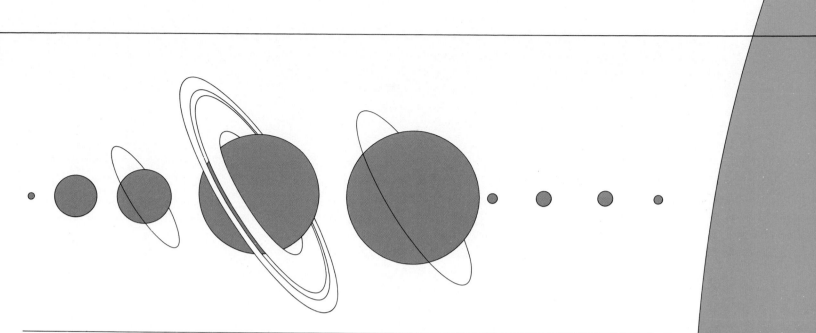

PLUTO	NEPTUNE	URANUS	SATURN	JUPITER	MARS	EARTH	VENUS	MERCURY	SUN
Charon	Triton	Miranda	Atlas	Metis	Phobos	Moon			
	Nereid	Ariel	1980S27	Adrastea	Deimos				
		Umbriel	1980S26	Amalthea					
		Titania	Janus	Thebe					
		Oberon	Epimetheus	Io					
			Mimas	Europa					
			Enceladus	Ganymede					
			Tethys	Callisto					
			Telesto	Leda					
			Calypso	Himalia					
			Dione	Lysithea					
			1980S6	Elara					
			Rhea	Ananke					
			Titan	Carme					
			Hyperion	Pasiphae					
			Iapetus	Sinope					
			Phoebe						

	MERCURY	VENUS	EARTH	MARS	JUPITER	SATURN	URANUS	NEPTUNE	PLUTO
Mean Distance from Sun (millions of miles)	36.0	67.1	92.9	141.5	483.4	886.7	1,782.7	2,794.3	3,666.1
Diameter (equatorial) (miles)	3,031	7,521	7,926	4,221	88,734	74,566	31,566	30,199	1,864
Mass (Earth = 1)	0.055	0.814	1.000	0.107	317.8	95.16	14.55	17.23	0.0026(?)
Density (water = 1)	5.43	5.24	5.52	3.93	1.33	0.71	1.31	1.77	1.1
Volume (Earth = 1)	0.06	0.86	1.00	0.15	1,323	752	64	54	0.01
Revolution around Sun	88.0 days	224.7 days	365.26 days	687.0 days	11.86 years	29.46 years	84.01 years	164.8 years	247.7 years
Rotation Period (days)	58.65	243.0	0.9973	1.0260	0.410	0.427	0.45	0.67	6.3867
Mean Orbital Speed (miles per second)	29.8	21.7	18.6	14.9	8.0	6.0	4.2	3.3	2.9
Inclination of Orbit to Earth's Orbital Plane	7.0	3.4	0.0	1.8	1.3	2.5	0.8	1.8	17.2
Gravity (Earth = 1)	0.38	0.90	1.00	0.38	2.53	1.07	0.92	1.19	0.05(?)

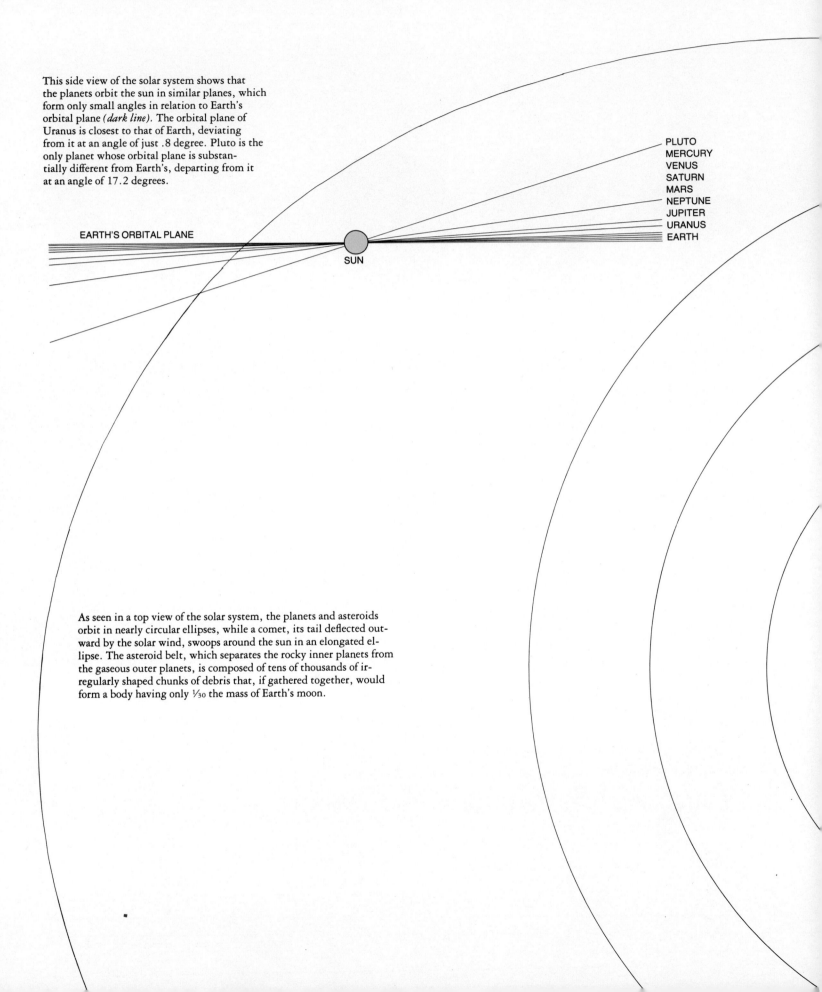

This side view of the solar system shows that the planets orbit the sun in similar planes, which form only small angles in relation to Earth's orbital plane *(dark line)*. The orbital plane of Uranus is closest to that of Earth, deviating from it at an angle of just .8 degree. Pluto is the only planet whose orbital plane is substantially different from Earth's, departing from it at an angle of 17.2 degrees.

PLUTO
MERCURY
VENUS
SATURN
MARS
NEPTUNE
JUPITER
URANUS
EARTH

EARTH'S ORBITAL PLANE

SUN

As seen in a top view of the solar system, the planets and asteroids orbit in nearly circular ellipses, while a comet, its tail deflected outward by the solar wind, swoops around the sun in an elongated ellipse. The asteroid belt, which separates the rocky inner planets from the gaseous outer planets, is composed of tens of thousands of irregularly shaped chunks of debris that, if gathered together, would form a body having only ⅟₃₀ the mass of Earth's moon.

The comet whose path is diagramed in the large drawing comes from a broad band of comets and other small bodies orbiting the sun at the outer edge of the solar system, as shown in the small diagram below. This band is called the Oort cloud after the Dutch astronomer, Jan H. Oort, who postulated its existence about one light year (5.87 trillion miles) from the sun. Comets begin their long journeys from the Oort cloud when they are nudged from their regular orbit by the gravity of nearby stars.

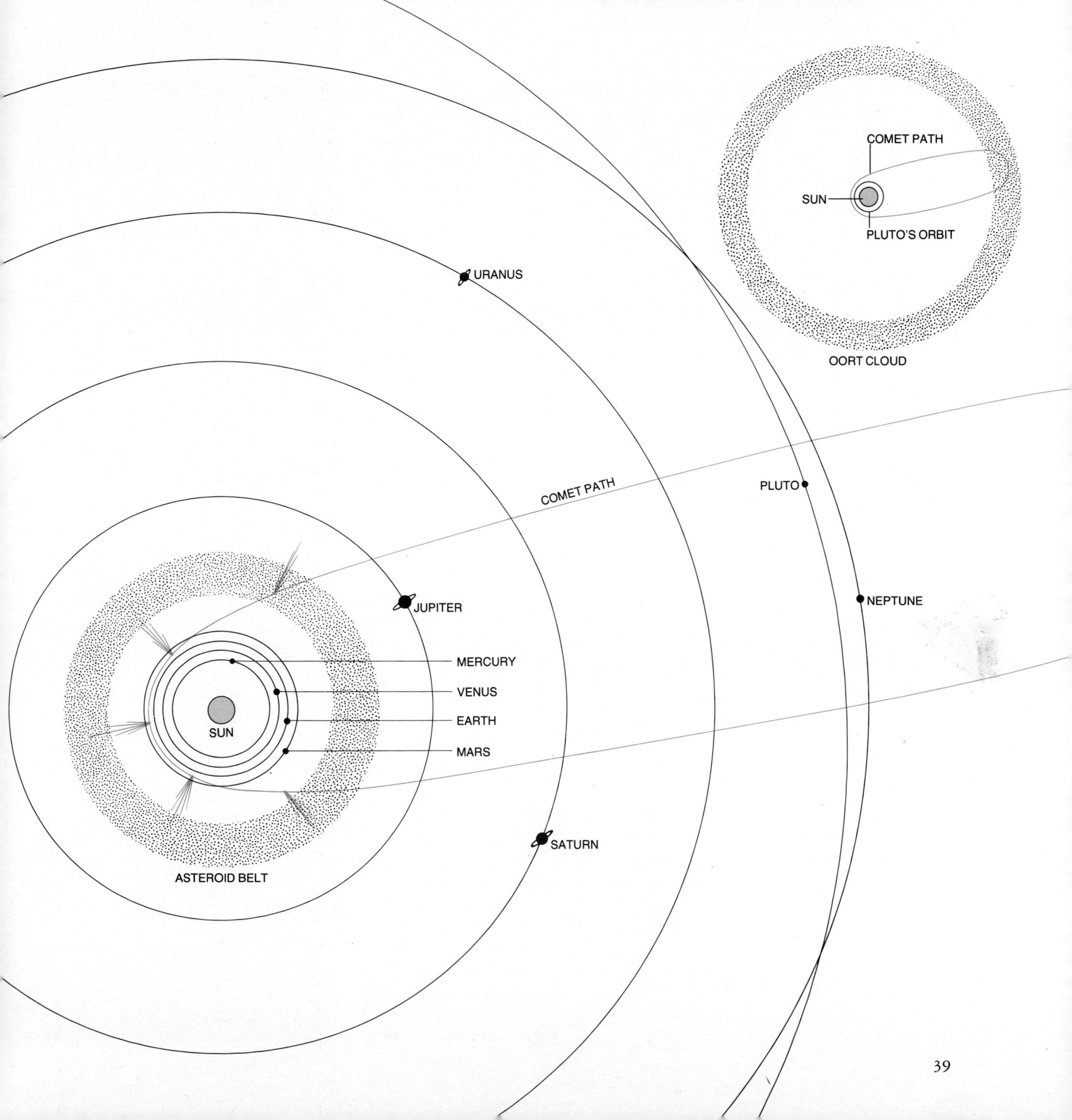

COMET PATH

SUN

PLUTO'S ORBIT

OORT CLOUD

URANUS

COMET PATH

PLUTO

NEPTUNE

JUPITER

MERCURY

VENUS

EARTH

MARS

SUN

SATURN

ASTEROID BELT

cal and electronic, they defined the exact shape of the closest major galaxy, a beautiful spiral named Andromeda, containing more than 300 billion stars. The nearest of these is 13 quintillion (13,000,000,000,000,000,000, or 13×10^{18}) miles, or 2.2 million light years, beyond the Milky Way, a distance calculated by comparing the star's apparent brightness with a star of similar brightness and known distance from Earth. And beyond Andromeda lay billions upon billions of other galaxies.

There were elliptical galaxies; barred spiral galaxies; galaxies turned broadside to show their grandeur in the fullest; galaxies revealed edge-on to profile their smooth, flattened disks and swollen galactic nuclei; irregularly shaped galaxies, sprawling haphazardly across intergalactic space; even colliding galaxies, two galaxies trying to pass through each other like enormous clouds sent headlong by strong winds, the unimaginable tidal forces distorting them into unnatural shapes. In the context of this immensity, the Milky Way, which was once thought to constitute the entire universe, now appeared as a tiny speck measuring a mere 100,000 light years across.

Advances in technology came more rapidly after World War II, ushering in the golden age of space exploration and repeatedly stunning the world with new insights into the dynamic workings of the solar system. Experimental research rockets poked their noses above the obscuring influence of Earth's protective atmosphere and permitted compact new electronic observational instruments to operate in the clear. And when in October of 1957 the Soviet Union launched *Sputnik,* the first man-made object put into orbit, the last impediments to virtually unlimited discovery finally seemed to be falling away.

The Russians and the Americans sent more-advanced space capsules on more-distant and difficult missions. In quick succession, spacecraft would measure the radiation belts around Earth and track global weather patterns. Still other spacecraft would record outpourings of X-rays, ultraviolet rays and gamma rays from the sun, providing clues to the inner workings of its nuclear furnace and to the behavior of flares that eject fast-moving particles throughout the solar system. Camera-equipped craft would land on the surface of the moon and fly by Venus and Mars, sending back details of surface features that, for centuries, had been little more than blurs at the end of a telescope lens. Then, suddenly, there would be men on the moon, human beings talking directly to Earth from another celestial body, collecting soil samples, recording moonquakes, and playfully making giant leaps and bounds in the weak lunar gravity.

The leap into space would continue right through the 1970s and into the 1980s. New generations of instrument-carrying spacecraft with names such as *Pioneer* and *Voyager* would transmit troves of eye-popping pictures from some of the distant gas giants. A project christened Viking would actually land two spacecraft on Mars with cameras and soil-sampling equipment. Monitoring the results of these probes, normally sedate researchers at NASA Mission Control whooped like schoolboys as the pictures from the planets flashed onto their screens. "Bizarre!" they shouted. "Wild!" "Incredible!"

It was all of that and a great deal more. For each new picture and each fresh piece of data would help to fill a gap in the ageless puzzle that is the cosmos. And the study of each body would bring a deeper understanding of how others evolved, and how they all fit together in the solar system. Ω

LIFE CYCLE OF THE SOLAR SYSTEM

About 4.6 billion years ago, a vast cloud of gas and cosmic dust began to collapse in a remote trailing edge of the Milky Way galaxy. Why it collapsed is one of the great mysteries of science. The probable cause was a prodigious explosion in the same galaxy. A gigantic star, no longer able to maintain its fiery bulk, had blown up in a supernova blast many times brighter than all the radiance of its 100 billion neighboring stars. Shock waves raced outward from the explosion and hit the cloud — and perhaps many clouds like it. Under the impact, the cloud began to contract and grow dense — a process that would eventually give birth to the solar system.

The sun and its nine planets were to be made up very largely of two elements, hydrogen and helium. These ingredients had been present throughout the universe since it was created in an expanding fireball 12 to 18 billion years earlier. The raw materials had coalesced into earlier generations of stars — generations that had perished and passed on matter to other spawning clouds. So too the sun and the planets would die at a predictable time and pass on their matter to future celestial bodies.

This startling scenario, once controversial but now generally accepted in scientific circles, was pieced together by scores of specialists after arduous and imaginative labors. They studied nearby stars of various known ages and established the general pattern by which stars develop and grow old. They marshaled voluminous new data on the evolution of the planets and their moons. Based on all this information, the paintings on these pages, by the noted science artist Don Davis, present the likeliest picture of how the solar system was born, and how it will one day perish.

In this time-lapse painting, a supernova explodes at top left, sending out shock waves. After traveling for tens of thousands of years, the shock waves reach a cloud of gas and dust, at right, and set up turbulence inside. In the inset painting of the Milky Way galaxy, a box shows the place where the solar system would emerge from the spawning cloud.

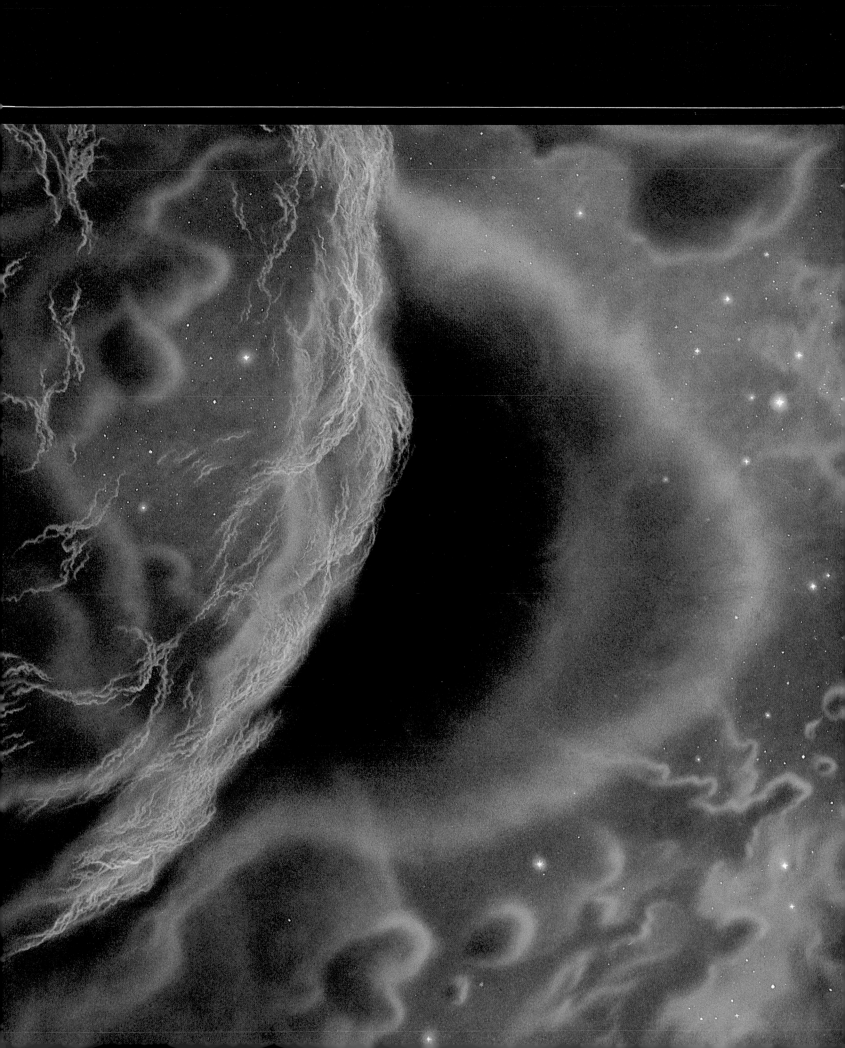

The Whirling Matrix of Sun and Planets

About 100,000 years after the spawning cloud began to collapse, it contracted into a swirling disk with a core that grew denser and hotter under the pressure of more and more gas and dust spiraling into the interior. This core was to evolve into the infant sun, and the nebula around it would form the planets and their moons. A preliminary phase was already under way: The different materials that would make up the inner and the outer planets were being sorted out by natural physical forces.

The heat of the core vaporized all but the rocky particles for 400 million miles around. These heavy particles would become the building blocks of the inner planets. Farther from the core, temperatures were cold enough for large volumes of water vapor and smaller amounts of other gases such as methane and ammonia to condense as ices and coat the relatively meager rocky particles. Thus the center of the outer planets would consist almost entirely of ice, which in time would attract large quantities of gases from the surrounding nebula.

Slowly, even as the sorting-out process continued, there began the actual build-up of particles that would create the planets and their moons.

As more and more matter is sucked into its center, a whirling cloud of gas and dust flattens into a disk-shaped solar nebula that spans nine billion miles, twice the extent of the fully evolved solar system.

New Worlds Built of Tiny Particles

The planet-forming process began gently and went on with increasing violence for millions of years. Like snowflakes, the particles in the solar nebula were fluffy and had a tendency to stick together on contact and form clumps. Under the influence of gravity, the clumps settled in a wide, thin sheet in the equatorial plane of the aborning sun. There the clumps began to stick to one another, building up into planetesimals — bodies up to several miles across. In turn the planetesimals collided with one another and coalesced into still larger bodies; the collisions became more violent as the bodies grew and gained appreciable gravitational force.

By this process of accretion, the planets were formed: four rocky inner planets — Mercury, Venus, Earth and Mars — and five icy or gaseous outer planets — Jupiter, Saturn, Uranus, Neptune and Pluto. The broad band of space between the inner planets and the outer ones was strewn with countless planetesimals possessing enough matter to form another planet. But the accretion process was thwarted by the gravitational pull of the giant Jupiter, which drove them into catastrophic collision rather than gentle coalescence. These planetesimals — now called asteroids — have therefore changed very little since they emerged with the solar system about 4.6 billion years ago.

The particles in the solar nebula settle into the sun's equatorial plane and begin combining with one another. In the inner reaches of the nebula, solid particles combine and build up mile-wide bodies of rocky material. In the cold outer regions, ice crystals and particles of frozen gases combine with dust to form orbiting chunks of dirty ice.

47

The End of
the Beginning

While dust and gas were accreting into planets, the dense, hot mass at the center of the nebula approached its final stage of development. The whirling gases continued to contract, exerting enormous pressure on the core and steadily raising the temperatures there. When the gases reached the critical temperature of 20 million degrees Fahrenheit, the hydrogen atoms fused, setting off a self-sustaining nuclear reaction. The sun was born.

The infant sun was not nearly as bright as it is today, but it was much more active. Huge flares erupted from the surface. Deadly ultraviolet rays — 1,000 times stronger than they are today — buffeted the planets. The solar wind — streams of particles from the sun's atmosphere — blew through space at speeds of about 2 million miles per hour, clearing the primordial solar system of the remaining gas and dust of the solar nebula. So powerful were the solar wind and the ultraviolet radiation that they stripped away much of the primitive atmospheres that had formed around the inner planets. The huge outer planets, protected by their greater distances from the sun and by their strong gravities, were able to retain their gaseous envelopes.

After about 9 million years, the solar wind abated. By then a new phase had begun for the planets and their moons.

The sun, its nuclear furnace newly ignited, sends out a fierce solar wind and powerful radiation that surge past the still-molten planets and blow away dust and gases left over from the formation of the solar system. In orbit around Earth, the infant moon *(foreground)* sweeps a path through debris too heavy to be blown away. Debris also forms a ring around Earth *(right),* which will gradually gather up the chunks. The solar wind has little effect on distant Jupiter and its forming moons *(far right).*

49

An Eon of Hurtling, Impacting Bodies

For hundreds of millions of years after the sun's birth, the solar system suffered the aftereffects of creation. On several of the planets and a few of their moons, volcanic eruptions relieved pent-up internal heat and ejected molten matter. And all of the planets and their moons were battered by remnants of the solar nebula that were too heavy to be swept away by the strong solar wind.

Hurtling bodies hit the planets in two distinct waves. The first onslaught consisted of small- to medium-sized chunks of the space debris that surrounded the planets. Then came a storm of icy planetesimals from the cold outer reaches beyond Uranus. These bodies invaded the inner solar system under the influence of the strong gravities of Jupiter and Saturn, which also flung planetesimals like stones from a slingshot far out into space. There they formed an immense cloud of comets that still orbits on the outer edge of the solar system.

The hurtling bodies were traveling at 20,000 to 50,000 miles per hour when they hit, and their tremendous impact carved out great depressions in the surface. Mercury, Mars and the moon still show these scars; on Earth, the craters were eventually eroded away or covered by seas and vegetation. Earth and perhaps Venus and Mars owe more than their craters to the debris from the outer solar system. As the icy bodies melted or vaporized in the heat of impact, they released their water, forming pools that would grow into oceans.

Chunks of space debris gouge out craters on Earth (*background*) and its moon. Earth, whose early ring has begun to shrink, is also buffeted by cyclonic winds, and its many erupting volcanoes belch out noxious gases and fiery orange lava. On the moon, the volcanic activity is not nearly as violent; lava gushing out of deep fissures floods craters and hardens to form smooth, dark fields.

50

Setting the Stage for Life on Earth

When the bombardment of space debris abated around 3.8 billion years ago, several planets and a few of their moons were still in the throes of intense volcanic activity. On Earth's moon, lava flows filled large impact craters. Volcanic vents discharged gases that, because the moon had little gravity, quickly escaped into space.

Earth with its stronger gravitational pull retained the gases accompanying volcanic eruptions. Carbon dioxide, water vapor, nitrogen, carbon monoxide and traces of ammonia and methane formed a thick atmosphere. Water condensed out of clouds and fell as warm rain over the cratered surface, enlarging the nascent oceans.

Into the warm waters fell molecules of amino acids, which were formed as the energy produced by the sun and by flashes of lightning combined atoms of gases present in the atmosphere. Amino acids were the building blocks of life; as free oxygen became available, complicated life-forms evolved.

That missing ingredient was first supplied more than 3 billion years ago, when primitive organisms that could tolerate the noxious air inhaled carbon dioxide and exhaled oxygen. Some of the oxygen was converted into ozone and built up a protective layer that shielded the planet from the sun's harmful ultraviolet radiation. The sun now became a nurturer rather than a destroyer of life.

Wracked by volcanic activity, the young Earth trembles in the final stages of formation. Gases released from volcanoes and from impacting meteorites feed the forming atmosphere and condense into violent thunderclouds over hot volcanic vents. The moon was then much closer to Earth, but the friction of tidal forces slowed Earth's rotation and rapidly moved the two bodies apart. They are still moving apart, though at the much reduced rate of about six inches a century.

A Bloated Sun, a Roasting Earth

As today's middle-aged sun grows older, its nuclear furnace will consume hydrogen at an accelerated rate, working harder and burning more fuel in order to maintain its internal temperature and pressure. As a result, the sun will slowly expand into a bigger, brighter, hotter star. It will shed more and more heat on the orbiting planets.

And slowly the climate of Earth will change. About 1.5 billion years from now, the temperate zones will receive their last snowfalls. The polar icecaps will melt, raising sea levels by hundreds of feet and inundating much of the land.

In another four billion years, when the sun is about 10 billion years old, it will have become almost 50 per cent larger than it is today. It will have exhausted most of its hydrogen fuel, and the nuclear fires will begin to die down. As the sun's core cools, its pressure will cease to be great enough to hold up the overlying solar layers, and the core will collapse.

The compression of matter in the core will again raise the temperature there, causing the surrounding layers to expand greatly. In the course of the next two billion years, the sun will balloon into a red giant, 100 times bigger and 500 times brighter than it is today. In its 12-billion-year-old enormity, the sun will engulf Mercury and raise daytime temperatures on Earth to 2,600° F. The once-lovely Earth will be seared to a lifeless crisp.

The bloated sun, its outline blurred by strong convection currents in the solar atmosphere, fills Earth's sky and melts its surface. The planet's atmosphere has long since been blown away by solar winds a million times stronger than they are today.

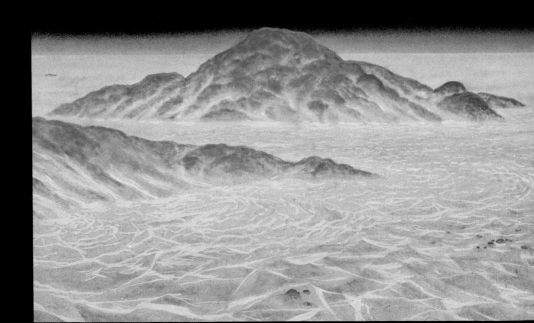

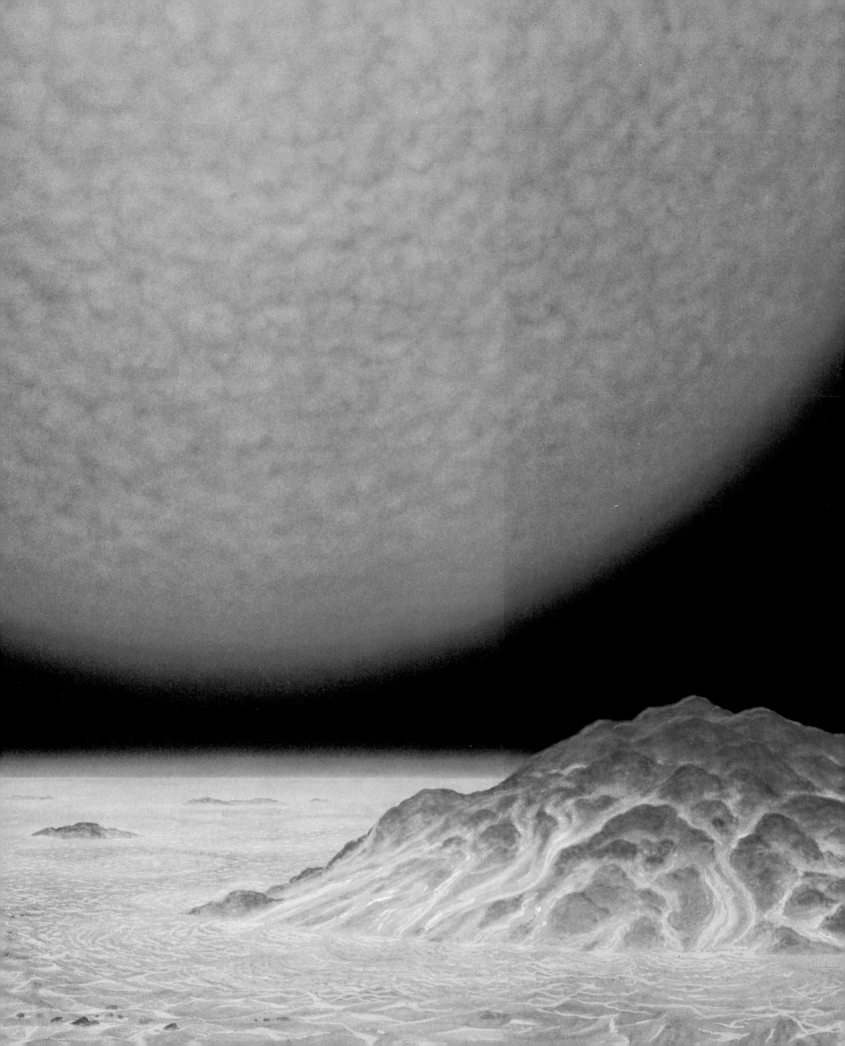

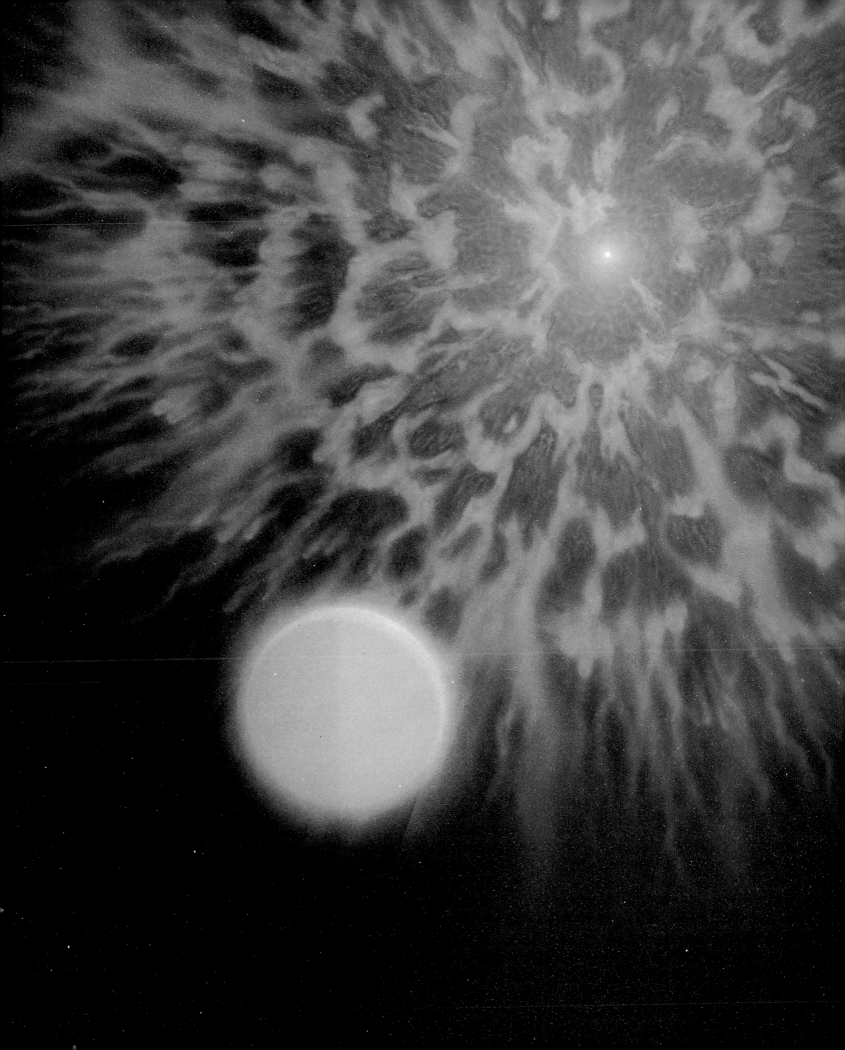

Into Dying Dwarf

The sun will survive as a red giant for about 250 million years. In this period, its core will fill with helium, produced by the nuclear fusion of hydrogen. The core will continue to contract and grow hotter. Eventually, the temperature will reach 200 million degrees Fahrenheit, hot enough to fuse helium atoms.

At this moment, the core will ignite in a helium flash, an explosive event that will last for several minutes. The core will expand violently, sending out shock waves that will expel as much as a third of the sun's mass.

Following the helium flash, the temperature of the core will cool slightly, enabling the sun to stabilize briefly as a helium-burning star. But then, the sun will again swell until the outer layers explode and are propelled far into space.

The sun is too small to disintegrate as a supernova like the one that probably triggered the formation of the solar system. Instead, it will be reduced to a small, bright, extremely dense star called a white dwarf; its remaining matter will be so densely packed that one teaspoonful would weigh five tons. With all of its fuel exhausted, the sun will gradually waste away and become a cold, dead body — a black dwarf. Yet even as the sun dies, the elements it produced throughout its life will survive somewhere in interstellar space and, mingling eventually with a spawning cloud of gas and dust, help to give birth to a new star, new planets and possibly even new life.

The sun's core ignites in a helium flash, ejecting the outer layers of solar matter; what is left of the sun will shrink into a small, bright white dwarf (*inset*). If Earth still exists then, it will be no more than a drifting cinder.

Chapter 2

THE SUN MACHINE

While scientists were steadily increasing their knowledge of celestial mechanics, they were learning very little about the most conspicuous feature in the solar system—the sun itself. How could earthbound observers, gazing across a gap of 93 million miles, manage to understand the forces that drove the fiery solar engine? When it came to the sun's physical structure and dynamics, astronomers at the turn of the 19th Century knew scarcely more than their Bronze Age counterparts at Stonehenge. In the absence of hard data, theory ran wild. As respected a figure as Sir William Herschel, knighted for his discovery of Uranus, seriously proposed in 1795 that the sun was a dark, solid body enveloped in luminous clouds and inhabited in its cooler regions by beings whose organs, he wrote, "are adapted to the peculiar circumstances of that vast globe."

Before long, however, the mysteries of the sun did begin to yield to persistent observation and to new discoveries in the fields of chemistry and physics. It became clear that the sun influences life on Earth more directly, and in many more ways, than astronomers had recently dreamed. Today, scientists recognize the sun as a medium-sized star of fascinating complexity, composed largely of hydrogen and helium, and powered by a thermonuclear furnace. In the depths of the solar core, hydrogen nuclei under the crushing pressure of the sun's mass are squeezed to form nuclei of helium. The energy released by this fusion of nuclei works its way slowly to the sun's roiling surface, then outward into the solar system at the speed of light. Sudden surges in the solar wind—the mighty stream of particles that flows through space and around Earth like a river around an island—can trigger geomagnetic storms that light up the skies and play worldwide havoc with power and communications systems.

This picture of a dynamic sun began to emerge with the work of Samuel Heinrich Schwabe, a German pharmacist-turned-astronomer, who commenced his solar observations in 1826. Initially he was looking for a new planet that might be orbiting close to the sun. But he soon gave up this quest to focus on sunspots, those dark smudges on the solar surface that had mystified astronomers since their discovery centuries earlier. As his obsession grew, Schwabe devoted more and more of his time to sun gazing, eventually selling the family drug business and abandoning himself entirely to his avocation.

Like the Copernican model of the solar system, the discovery of sunspots had been fraught with disturbing implications. They appeared to be blemishes on the solar disk, but any early astronomer who so described them

In a computer-enhanced photograph of a total solar eclipse, the black sphere of the moon hides the entire disk of the sun and is haloed by the sun's dazzling corona. The colors of the corona reflect the intensity of light: White is the highest, deep violet the lowest.

S. P. Langley, Del.

risked censure for challenging the orthodox Christian view of the sun as symbol of perfection. One of the discoverers of sunspots, a German Jesuit and mathematician named Christoph Scheiner, was forbidden by his ecclesiastical superiors to publish his findings in his own name, and Galileo delayed his announcement out of prudent concern for Church reaction.

Nevertheless, sunspots proved to be wonderful objects for study and speculation. With a telescope the sun's image could be projected onto a screen and the sunspots tracked on a daily basis. Like dark chimeras they danced across the surface—changing size and shape, forming into clusters, breaking apart and regrouping into long chains. From their movement, Galileo correctly inferred that the sun rotates on its axis like Earth. Other observers put forth various theories to explain these curious features. Some thought the sunspots might be clouds floating above the solar surface. Others suggested they were layers of slag that had accumulated from volcanic eruptions, or mountains floating in a molten sea. Herschel believed they were openings in the sun's cloud cover that allowed him to peek at the solid surface underneath.

Perhaps the oddest phenomenon was that the sunspots waxed and waned over long periods. They had virtually disappeared from the sun's surface between 1645 and 1715—a hiatus that investigators would later connect with a prolonged change in climate known as the Little Ice Age. In that 70-year period, temperatures in Europe averaged fully 2° F. lower than those in the several decades preceding and following.

Like others before him, Schwabe had noticed this variability in sunspots, but he went one step further to show that the changes followed a cyclic

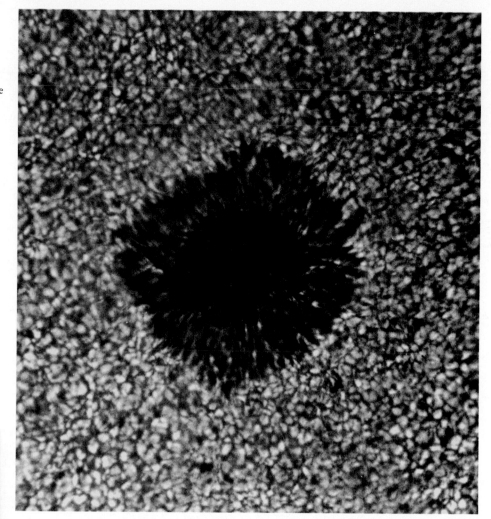

A recent photograph taken through a high-powered telescope shows a sunspot — dark in the center with a lighter fringe — surrounded by the cell-like granules that cover the surface of the sun. Though the sunspot looks unimpressive in size and luminosity, it was twice the diameter of Earth and 10 times brighter than Earth's full moon.

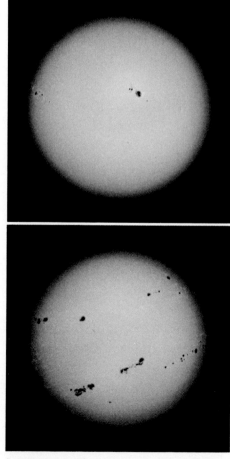

Sunspots wax and wane in an 11-year cycle. Typically, the solar surface shows only a few sunspots near the start of a cycle (top). Five years and a few months later, sunspot activity reaches its peak (bottom) — the solar maximum.

pattern. Schwabe was nothing if not methodical. He observed the sun on almost every clear day for 17 years, culminating remarkably with 312 days in 1843. His records demonstrated a wavelike trend in the number of sunspot groups. From a peak of 225 in 1828, the clusters receded to a low of 33 in 1833. Then the numbers began climbing again, reaching a new high of 333 groups in 1837 before dropping to a minimum of 34 in 1843. This tenfold difference between peaks and valleys occurred approximately every five years, so that a complete period in the cycle spanned about a decade. Clearly the sun danced to some sort of internal metronome, keeping time with its own stellar rhythm.

Schwabe's dramatic revelations excited tremendous interest in sunspots. As scientists accumulated longer solar records, they gradually refined Schwabe's calculations; today the sunspot cycle is recognized as averaging 11.2 years. Astronomers have also determined that sunspots are areas of intense magnetic activity and that their darkness results from a contrast in temperature: They are about 3,600° F. cooler than the white-hot surrounding gases, which reach 11,000° F.

Among the generation of sun watchers who followed Schwabe was Richard Christopher Carrington, a young Englishman of independent means who spent eight years mapping sunspots at a private observatory in Surrey. Carrington's persistence paid off in new scientific laws about sunspot distribution and solar rotation. But his most startling discovery was made in a flash. At 11:18 a.m. on the morning of September 1, 1859, he witnessed a tremendous eruption on the surface of the sun.

He later recalled the phenomenon in detail. "Two patches of intensely bright and white light broke out. I thereupon noted down the time by the chronometer, and seeing the outburst to be very rapidly on the increase, and being somewhat flurried by the surprise, I hastily ran to call someone to witness the exhibition with me, and on returning within 60 seconds was mortified to find that it was already much changed and enfeebled. Very shortly afterwards the last trace was gone, and although I maintained a strict watch for nearly an hour no recurrence took place."

Carrington was the first person to witness and report a solar flare. His observation would have been important enough in itself. But the next day a series of extraordinary events was recorded around the world — events that, in Carrington's mind, might have been linked with the explosion of light he had seen on the sun. At 4 a.m., electrical currents were induced in telegraph lines so that messages could be sent along them without batteries. That night the northern lights, or aurora borealis, put on a spectacular display across Europe and America, extending far south of their normal range. Westering pioneers camped beneath the lights on the Kansas prairie, while residents of Honolulu — only 21° north of the Equator — gazed in utter astonishment at their shimmering brilliance. The worldwide magnetic storm also set off a flamboyant display of the aurora australis in the Southern Hemisphere.

Showing proper scientific caution, Carrington avoided claiming that there was a cause-and-effect relationship between the solar flare he had witnessed and the bizarre happenings on Earth. "One swallow does not make a summer," he said. Nevertheless, this was one of the most dramatic demonstrations ever recorded of an apparent link between violent activity on the sun and events on Earth.

Over the next century, solar flares periodically visited their wrath on the planet. But it was not until scientists developed Earth-orbiting spacecraft equipped with X-ray telescopes and other space-age instruments that they were able to observe these eruptions in great detail and explain them comprehensively.

Like sunspots, the flares are associated with magnetically active regions of the solar surface. In a typical flare, energy builds up in an arcing magnetic loop that becomes increasingly unstable, until it finally erupts. In the resulting explosion, powerful X-rays, gamma rays and high-energy particles are spewed into space, buffeting Earth, disrupting the planet's magnetic field and triggering geomagnetic storms. This rain of particles induces currents in power cables in the ground and along pipelines. The voltage changes can interrupt long-distance power transmissions, damage electrical equipment and affect military radar. A flare may create disturbed conditions in the upper atmosphere; as a result, short-wave transmissions are absorbed and radio communications blacked out. Flares can affect more than human technology. The magnetic storms associated with one solar flare disrupted a homing-pigeon race, disorienting almost all the birds by impairing their acute sensitivity to Earth's magnetic field.

A flare's burst of energetic particles also accounts for the spectacular surge in the northern lights. Auroras result when the magnetic fields of Earth and the sun interact to generate electricity. The sun's magnetic field rides on the stream of energized particles known as the solar wind — a wind that becomes a raging hurricane when a flare erupts, lashing Earth with high-energy particles. The powerful current induced in Earth's mag-

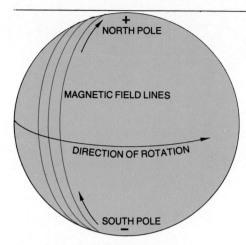

At the beginning of the sun's magnetic cycle, the magnetic field lines run north-south between the poles, one of them with a positive polarity, the other negative.

Invisible Lines of Force

Sunspots and other solar phenomena are visible evidence of an invisible internal force: the sun's magnetic field. The force is believed to stem from circulating electrical currents in the outer third of the solar sphere.

The rotation of the sun shapes the magnetic field. Since the sun rotates faster at the equator than at the poles, the sun's magnetic lines of force are slowly stretched like elastic bands and pulled laterally from their basic pole-to-pole axis. Simultaneously, the lines rise and fall along with currents of hot gas and are twisted and intertwined into magnetic ropes.

The stretching and twisting increases the strength of the magnetic field; the lines of magnetic force squeeze out gases between their strands and rise to the surface. There they accumulate, forming sunspots and releasing energy as solar flares and other eruptions.

This varied solar activity reaches a peak every 11 years. Then the magnetic lines begin to unwind, and the sun returns to its state of lowest magnetism — and least activity.

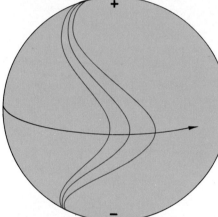

The sun's rotation (arrow) causes the magnetic field lines to start stretching. The effect is most obvious at the equator, which rotates faster than the higher latitudes.

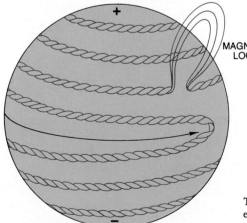

The magnetic field lines continue to stretch, wrapping themselves around the sun. They are also twisted and buoyed upward. Eventually they burst through the surface and form loops.

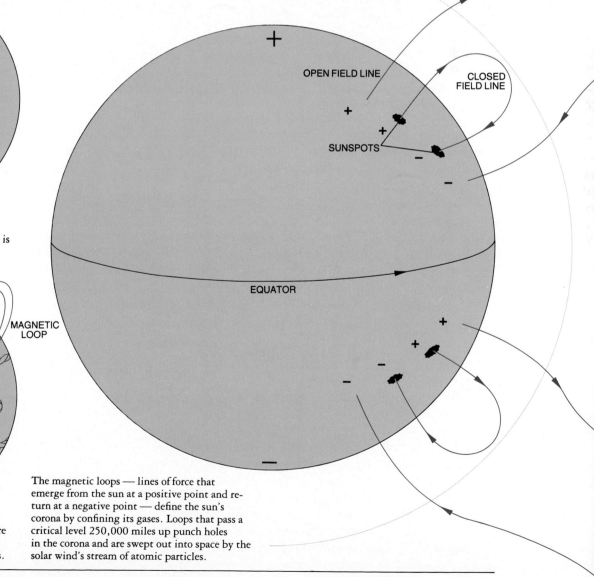

The magnetic loops — lines of force that emerge from the sun at a positive point and return at a negative point — define the sun's corona by confining its gases. Loops that pass a critical level 250,000 miles up punch holes in the corona and are swept out into space by the solar wind's stream of atomic particles.

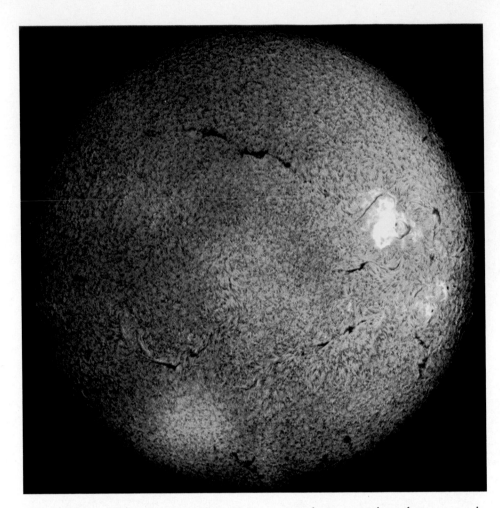

A photograph taken through a special hydrogen alpha filter, which isolates part of the visible spectrum, reveals one of the largest solar flares recorded, seen as a bright patch at right. This vast eruption, which took place on August 7, 1972, covered a solar area equal to 10 Earths, and the charged particles it flung earthward caused widespread power blackouts and telecommunications failures.

netic field excites atoms in the upper atmosphere, causing them to emit light in certain characteristic colors — green for oxygen, crimson for nitrogen. When geomagnetic storms are particularly intense, they distort Earth's magnetic field, extending the range of auroras to lower latitudes where they are not normally seen.

With the arsenal of sophisticated devices available to them, modern astronomers maintain a constant watch on the sun, and they often are able to predict the outbreak of solar flares and the geomagnetic storms that follow. The biggest such event of recent times took place in August of 1972. Solar observers had been tracking a rather ordinary sunspot group and had watched it pass out of view as a result of the sun's rotation; their suspicions were aroused two weeks later when the sunspot group reemerged appearing utterly different. The group was organized in a dense and complex structure, and a pair of sunspots had grown significantly in size. Scientists at California's Mount Wilson Observatory reacted quickly. With special sensors they monitored the magnetic fields of the region and were shocked at their intensity.

Solar scientists braced for a major eruption. In the course of a 15-hour period on August 2 the region cut loose with three gigantic flares that ejected matter that accelerated to speeds of as much as 2.7 million miles per hour — tripling the velocity of the solar wind. The solar eruptions would continue until August 7, with the largest flares covering nearly 80,000 square miles of the sun's surface.

The first to feel the effects was the *Pioneer 9* spacecraft, orbiting the sun between Venus and Earth. Thirty-three hours after the first flare, the tiny craft passed through an interplanetary shock wave. Solar forecasters at the Space Environment Services Center in Boulder, Colorado, issued a warning

The aurora borealis spreads a curtain of shimmering light across the night sky near Fairbanks, Alaska. The northern lights and their southern counterpart occur because the solar wind collides with Earth's magnetic field, transferring energy that causes the upper atmosphere to glow around the Poles. The auroras are brighter and range farther over the globe after a strong solar flare.

that Earth would be hit by the first of several geomagnetic storms during the night of August 3.

During the next several nights a brilliant aurora radiated over much of North America and Europe. Power lines surged with voltage fluctuations, and a 230,000-volt transformer in British Columbia exploded. The storm tripped circuit breakers and damaged electrical filters in long-distance telephone systems. Headlines around the world trumpeted the news: "The Sun Erupts." Later, a scientific report would refer to these storms' impact as "geophysical events of historic proportions." Yet the storms were actually less intense than some during the previous century. They were just better documented and their effects more widespread in an age of power grids and global telecommunications.

Flares are but one of several events causing surges in the solar wind. Solar prominences — great loops of plasma, or hot gas, that may soar more than 300,000 miles into space — can produce similar effects and are no less dramatic. There are two main types of prominences: quiescent and active. Quiescent prominences appear as bright arches at the edge of the sun or as dark ribbon-like filaments extending up to 125,000 miles across the sun's disk. They may last a few days or many months, but they rarely trigger electromagnetic effects on Earth. On the other hand, active prominences last only a few hours, can erupt into space (at speeds of up to 2.4 million

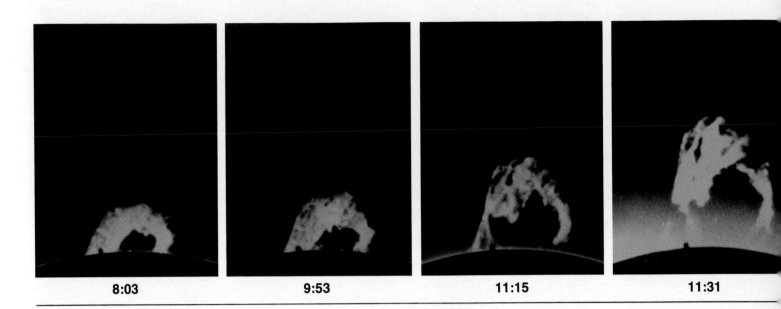

| 8:03 | 9:53 | 11:15 | 11:31 |

miles per hour) and often produce effects on Earth. Both types of prominences are denser and cooler than the sun's corona, or outer atmosphere.

Flares and active prominences may trigger huge shock waves in the sun's corona, forming coronal transients. These formations appear as magnetic loops or bubbles thrown out from the sun. They were first observed by the satellite OSO-7 on December 14, 1971, then photographed extensively from the Skylab manned space mission in 1973.

Skylab also helped to resolve a long-standing mystery about other geomagnetic disruptions that were apparently solar in origin. The disturbances occurred at roughly 27-day intervals — the period of the sun's rotation as seen from Earth — and usually involved an interruption of radio transmissions. Scientists guessed that the source of these disruptive discharges must be hidden in some active regions that do not reveal themselves in the usual way through sunspots, flares or prominences. In 1932 a geophysicist coined the term M-regions to describe these areas, with the *M* standing for both "magnetically active" and "mysterious."

Scientists remained baffled by M-regions for almost 40 years. At last, a clue to their identity was found in 1970; photographs taken during a solar eclipse revealed a dark gap in the corona, which is normally obscured by the brightness of the sun's disk and can only be seen during an eclipse. It was known that the corona partly restricts the outflow of particles constituting the solar wind. It seemed possible, then, that the gap in the corona could funnel into the solar wind a concentrated stream of particles, creating a powerful current in the interplanetary ocean.

A critical piece of the puzzle fell into place when scientists in the early 1970s examined some startling X-ray photographs of the sun. These pictures were taken from rockets fired above Earth's atmosphere, which blocks out X-ray emissions, and they revealed what appeared to be holes in the corona. Moreover, magnetographs — instruments that record the polarity and strength of magnetic fields on the sun — showed that the holes were acting as escape hatches for the sun's magnetic force lines.

That result was confirmed in 1973 by the American solar physicist Allen S. Krieger. He succeeded in tracing back to its point of origin on the sun a stream of high-velocity particles bombarding Earth. Krieger was able to

A steady eruption of flaming gases forms a solar prominence, shown in photographic sequence as it waxes and wanes over a four-hour period from 8:03 a.m. to 12:10 p.m. on August 18, 1980. Some solar prominences are more stable and persist for weeks or months.

66

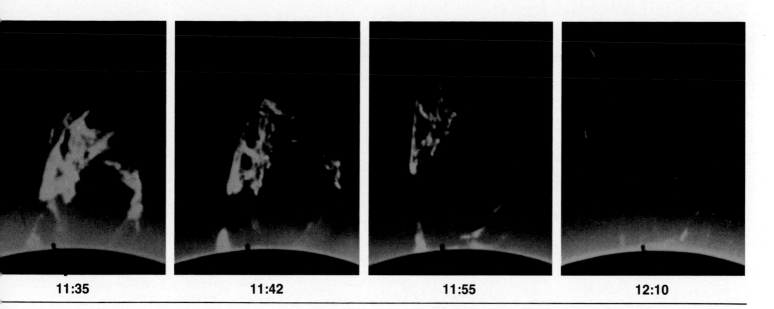

| 11:35 | 11:42 | 11:55 | 12:10 |

show unequivocally that the particles were shooting out of a coronal hole.

Meanwhile, the Skylab mission began compiling an extraordinary nine-month record of X-ray data on coronal holes and other solar phenomena. A succession of Skylab crews shuttled between Earth and their orbiting home in space, and about 30 per cent of their waking time aloft was spent sun watching at the console of Skylab's ATM (Apollo Telescope Mount). The astronauts became obsessed by the great solar disk looming on the TV screens, and they would have spent many more hours observing it if Mission Control had allowed. The sun's granular surface was enhanced by viewing it at certain wavelengths. To one crew member it resembled "a great big bowl of oatmeal with pepper on it."

The data collected by the Skylab astronauts finally laid to rest the mystery of the M-regions. A definite correlation was shown to exist between the coronal holes and geomagnetic events recorded by earthbound instruments. M-regions are simply gaps in the corona — floodgates for the stream of solar particles washing over Earth.

These discoveries left unsettled a long debate concerning variation in the amount of solar energy reaching Earth. Was there in fact a much-discussed "solar constant," or was the sun's energy output erratic? The question was of more than academic interest, for even small energy changes can, in time, have potentially catastrophic effects on Earth's climate and food production.

For a good part of the 20th Century an indefatigable scientist at the Smithsonian Institution in Washington, D.C., sought to prove that the solar constant was in fact inconstant — that the sun was a "variable star" whose energy output changed in predictable cycles, albeit less noticeably than that of other stars. Charles Greeley Abbot believed that the amount of solar energy reaching Earth varies and that the variation has had demonstrable effects on climate. In 1902 he began systematically collecting data on solar radiation from stations on four continents. Abbot's work continued until 1953 and seemingly documented changes between .1 and 1 per cent in the amount of solar energy reaching Earth. These figures neatly bracketed climatologists' estimates that a .5 per cent variation in the solar constant could alter weather patterns on Earth.

In the Shadow of the Moon

Eclipses of the sun and moon have long been described in myths as the work of fierce gods or voracious dragons. But they might better be characterized as nature's most spectacular parlor tricks. An eclipse occurs only when Earth, its moon and the sun are in a straight line, with Earth or the moon blocking the sun. The positions of the bodies during the three types of eclipse are diagramed below.

A total solar eclipse is the most sensational type. The moon is ¹⁄₄₀₀ the size of the sun, and in order to cover the solar disk it must pass close to Earth in its elliptical orbit. When the moon is properly aligned, a partial eclipse can be seen from within the moon's penumbra, or partial shadow. As seen from inside the moon's umbra, or main shadow, the whole solar disk is briefly blotted out.

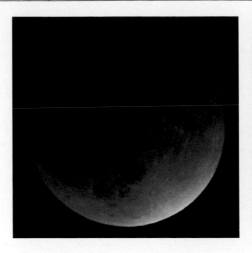

During a lunar eclipse, the moon is illuminated by stray rays of sunlight that bend around Earth as they pass through the atmosphere.

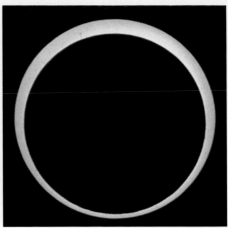

A solar halo rings the moon in an annular eclipse of the sun. In such eclipses, the moon is too far from Earth to hide the sun completely.

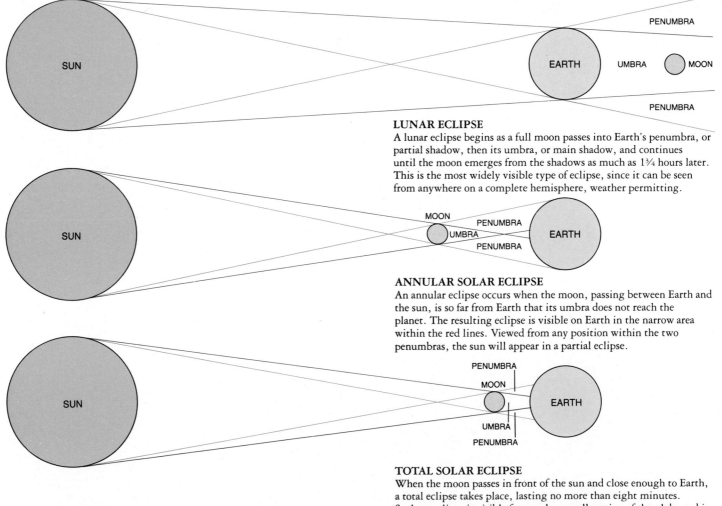

LUNAR ECLIPSE
A lunar eclipse begins as a full moon passes into Earth's penumbra, or partial shadow, then its umbra, or main shadow, and continues until the moon emerges from the shadows as much as 1¾ hours later. This is the most widely visible type of eclipse, since it can be seen from anywhere on a complete hemisphere, weather permitting.

ANNULAR SOLAR ECLIPSE
An annular eclipse occurs when the moon, passing between Earth and the sun, is so far from Earth that its umbra does not reach the planet. The resulting eclipse is visible on Earth in the narrow area within the red lines. Viewed from any position within the two penumbras, the sun will appear in a partial eclipse.

TOTAL SOLAR ECLIPSE
When the moon passes in front of the sun and close enough to Earth, a total eclipse takes place, lasting no more than eight minutes. Such an eclipse is visible from only a small section of the globe and is the rarest type of eclipse. In adjacent areas it appears as a partial eclipse, and elsewhere it cannot be seen at all.

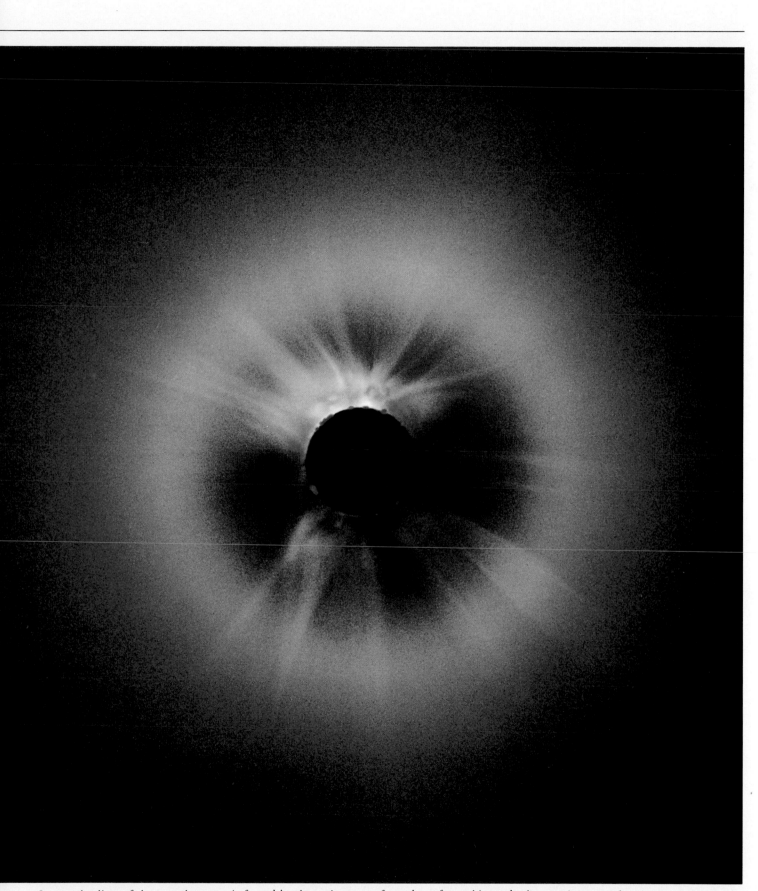

In a total eclipse of the sun, the moon is framed by the sun's gaseous fireworks — far-reaching red solar prominences and white coronal streamers.

But most solar scientists were not convinced. In a reassessment of Abbot's work following his death in 1973 — he lived to the age of 101 — his colleagues concluded that he had innocently attributed a false precision to his data. The problem came from trying to record radiation filtered through Earth's atmosphere, whose opacity varies according to weather and the time of year.

That was not the end of the story, however. Seven years after Abbot died, NASA's unmanned Solar Maximum Mission — the orbiting successor to Skylab — began collecting data on the solar constant from above the atmosphere. In 1980, Richard C. Willson of the Jet Propulsion Laboratory reported on the first five months of observations. The satellite data indicated that the sun's outpouring of energy varied almost daily. The amounts of change were generally small, about .05 per cent, or 1 part in 2,000. But there were two much larger dips of about .2 per cent, or 1 part in 500.

These dips were intriguing in another way. Both occurred just as a large group of sunspots was passing across the central part of the sun's disk, suggesting that the sunspots were blocking some of the solar radiation. Observations over the next year continued to show frequent variations in solar output, with one drop of about .23 per cent — the deepest dip yet measured — again coinciding with the passage of a major sunspot group. Willson proposed that the sunspots might somehow be impeding the radiation of energy from the solar surface. In fact, the observed energy output decreased by exactly the amount estimated from the total sunspot area. He postulated that the "missing energy" is stored for several months and then leaks out over a large region.

These suggestions prompted lively disagreement among solar physicists. A counterexplanation held that the missing energy is radiated away horizontally at some distance from the sunspots through magnetically active, bright patches known as faculae.

Whatever caused the energy variation, there could be no doubt that it was occurring. Besides the small changes over a few days or weeks, the data showed a decrease in solar output of about .1 per cent over the first 18 months of the mission, after which the output began rising again. The Solar Maximum Mission had proved irrefutably that the solar constant is not constant after all. It had vindicated Charles Greeley Abbot's lifetime effort to demonstrate that the sun is, in an important sense, a variable star.

The composition of the sun was still a matter of baseless speculation in the middle of the 19th Century; indeed, a school of empirical philosophers declared that such information lay beyond the bounds of reasonable scientific inquiry. But then a pair of clever German scientists revolutionized astronomy with development of the recording spectroscope, an instrument that allowed them to decipher the code of the solar spectrum and learn the chemical make-up of the stars.

Chemist Robert Bunsen, who gave his name to the Bunsen burner used in laboratories, and physicist Gustav Kirchhoff began their collaboration in 1854. They discovered that each element, if it is heated to incandescence and its lights diffracted into a spectrum, produces a characteristic "signature." Pure sodium, for example, showed as a bright double yellow line.

Using Bunsen's new burner, which gave off a very high-temperature flame, the two scientists began cataloguing elements based on their spectral

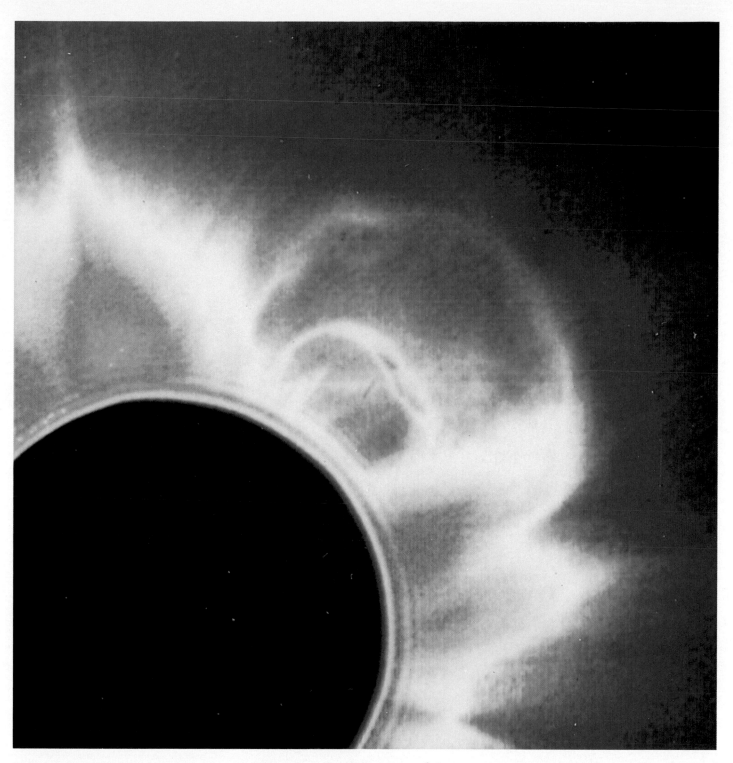

Triggered by a powerful eruption of solar particles, a double-arched coronal transient balloons out into space. Some transients have been estimated at one million miles across — a distance greater than the sun's diameter.

signatures. In the process of these systematic experiments, the men noticed that the bright signature line produced by a high-temperature flame was reversed — that is, converted to a dark line — if the light was passed through a low-temperature flame from the same element. It was apparent that the low-temperature flame absorbed the bright signature line of the high-temperature flame.

Kirchhoff then made a crucial connection — one that would prove as important to astronomy as the invention of the telescope 250 years before. Since the early part of the century, scientists had noticed that the sun's spectrum was scored at many places by dark lines. Now, in an exhilarating flash of insight, Kirchhoff concluded that these mysterious features were in fact the signature lines of elements present in the sun's atmosphere. The atmosphere, being relatively cooler than the sun's interior, absorbs certain wavelengths of light and renders them as dark lines in the spectrum.

Bunsen conveyed his excitement in a letter to a colleague in England: "At present Kirchhoff and I are engaged in an investigation that doesn't let us sleep. Kirchhoff has made a wonderful, entirely unexpected discovery in finding the cause of the dark lines in the solar spectrum. Thus a means has been found to determine the composition of the sun and fixed stars."

Using a special instrument that combined a spectroscope with a pair of superbly crafted telescopes, Kirchhoff could see that the solar spectrum contains thousands of lines, each remarkably different in breadth and position, and organized in a variety of groupings. He proceeded to chart the solar spectrum, drawing the lines with utmost precision. His instrument allowed him to compare the sun's spectrum with the spectrum of a source of incandescent light in the laboratory. The rays from one source were led through the upper half of a vertical slit and matched against the rays of the other source projected through the lower half of the slit. The two spectra were thus directly aligned, one under the other. It was easy to see whether any lines coincided.

"I have in this way assured myself that all the bright lines characteristic of iron correspond to dark lines in the solar spectrum," Kirchhoff exulted. He also found the lines of magnesium, chromium and nickel. The chemical analysis of the sun was under way.

The dark lines in the solar spectrum are called Fraunhofer lines for an earlier German physicist who had discovered and catalogued them in 1814. Joseph von Fraunhofer had designated each line grouping by a letter. Using the marvelous new method of Kirchhoff and Bunsen, scientists were able to identify the element represented by almost every line grouping in the Fraunhofer catalogue. They found that the solar groups known as C and F lines identified the element hydrogen. H and K were the signatures of calcium. And so on.

One line, however, resisted all laboratory identification. It was first noticed in the spectrum of a solar prominence during an eclipse in 1868. The two D lines associated with sodium in the yellow-orange part of the spectrum were there, along with another line — described as "near D" — just a short distance to the right. The French astronomer who observed the eclipse reported the unknown line to a young British solar astronomer named Norman Lockyer, who went to work immediately trying to identify it.

Lockyer, who eventually was knighted for his work in solar spectroscopy, made his own observations of the mysterious line and conducted experiments to see if he could reproduce it in the laboratory. But the line — designated D^3 to distinguish it from the sodium lines D^1 and D^2 — did not yield its secrets easily. Lockyer wrote: "It was found that there was no substance in our laboratories which could produce it for us, whereas in the case of the line D we simply had to burn some sodium, or even common salt, to produce it."

Further studies showed that the position of the line moved from side to side — representing a slight change in observed wavelength — from one solar observation to the next. Apart from the mystery of the line's identity, Lockyer realized that this wavering was due to the motion of gases in the sun's atmosphere either toward or away from the observer. This fluctuation was the familiar Doppler shift, the same phenomenon that causes

Calcium

Hydrogen

Calcium

Iron
Hydrogen
Iron

Magnesium

Hydrogen

Iron

Magnesium

Iron
Iron

Chromium

Iron

Calcium
Sodium

Calcium

Iron

Hydrogen

Nickel

72

This spectrum — a rainbow-hued bar crossed at telltale intervals by dark lines — reveals the sun's chemical make-up. The colors represent light emitted by the sun's surface layer, the photosphere. However, certain wavelengths of light are absorbed by the gases in the sun's atmosphere and therefore appear as dark lines. Moreover, some elements may exist in more than one energy state and thus have more than one line. The solar spectrum shown here is much abbreviated; the full spectrum contains thousands of lines and extends more than 40 feet.

the change in pitch of a whistle on a passing train; its sound waves are compressed, then elongated, as the train approaches and recedes from an observer. The shift observed in the D^3 line of the solar spectrum enabled Lockyer to calculate that its movements on the sun reached, and sometimes exceeded, the amazing speed of 100 miles a second. It also showed him that D^3, which in some ways was acting like hydrogen, was not some new line of hydrogen, because the known Fraunhofer lines for hydrogen did not change position at the same time.

Lockyer then concluded "that we were not dealing with hydrogen; hence we had to do with an element which we could not get in our laboratories, and therefore I took upon myself the responsibility of coining the word helium, in the first instance for laboratory use." He had taken helium from the Greek word for the sun, *helios.* But helium itself would not be found in any terrestrial material until 27 years later. It was discovered in 1895 as an inert gas emitted in the laboratory from a type of uranium ore called cleveite. The element was later found as a minor constituent in natural gases associated with oil wells.

Scientists have learned that hydrogen and helium together account for nearly all the atoms in the sun — 99.9 per cent. In the tiny remaining fraction are 80 of the other 90 natural elements. Hydrogen and helium constitute, respectively, 92.1 per cent and 7.8 per cent of the sun's atoms — a ratio of more than 11 to 1. Because of differences in their atomic weight, the ratio of hydrogen to helium in terms of the sun's total mass is about 3 to 1. Thus, just under 75 per cent of the solar mass is in the form of hydrogen, with helium making up almost all of the remaining 25 per cent.

Spectroscopy opened a door on solar science, yielding priceless data not only on the sun's chemical make-up but on its temperature, density, magnetism and surface dynamics as well. The spectrograph, along with the spectroscope and other forms of the instrument, enabled scientists to compare the sun with other stars *(page 75)* and to work out astonishingly detailed theories about the birth and evolution of stars, galaxies and the universe itself. Leaders in several specialized fields contributed to these advances. But more than anyone else, an American named George Ellery Hale was responsible for developing the essential tools for unlocking the awesome secrets of the sun.

Hale was an imaginative astrophysicist who approached his studies as an experimental scientist at a time when most astronomers were less interested in the stars' physical properties than their position, distance and motion. As early as 1889, at the age of 21, he devised a type of spectrograph that he called a spectroheliograph. This apparatus could be focused down to admit the spectral light of only one element, filtering out brilliant light that had obstructed study of the sun's outer layers.

At California's Mount Wilson Observatory — one of three major observatories he was instrumental in founding — Hale and his associates took the first photograph of a sunspot spectrum in 1905; their observations confirmed Hale's long-standing suspicion that sunspots are cooler than other solar areas. Three years later, using photographic plates sensitive to red light, Hale detected vortexes in the glowing jets of hot gas near sunspots, leading him to postulate that intense magnetic fields were associated with sunspots. Then, using a powerful electromagnet, he duplicated in the laboratory certain effects of magnetic fields on sunspot spectra, proving for the

Discovering the Make-up of Stars

A versatile tool in analyzing stars, spectroscopy reveals not only a star's chemical make-up but also its color and temperature. Spectroscopy is also used with various formulas to determine a star's mass, size and luminosity, or brightness. By comparing a star's temperature with its brightness, scientists can place the star in one of four categories according to its stage of evolution.

Most stars are middle-aged, small to medium in mass and size, and burn their hydrogen fuel at a stable rate; they are known as main sequence stars. The smallest, least massive of these stars are red; they have less fuel to burn and thus are the dimmest and coolest in the category. The largest, most massive ones have more fuel and burn it faster, mak-

ing them the brightest and hottest main sequence stars. Their color is blue. The sun is a relatively small, dim and cool main sequence star. It is yellow-orange.

Unlike main sequence stars, stars in the three other categories are not consistent in their relationship between size, temperature and brightness. Giants and supergiants, now in the unstable later stages of their evolution, increase enormously in size and brightness as they age. But as their outer shells swell, their brightness is spread over a much larger surface area, making them relatively cool and therefore red. Stars of the fourth category are known as white dwarfs. Each is the dying core of a giant that has exhausted its fuel and expanded until it burst, shedding its outer layers.

TEMPERATURE AND BRIGHTNESS
This chart shows how brightness and temperature vary in the four categories of stars. Alnilam, at top left, is a very bright, very hot supergiant. The six stars below it on the descending diagonal are main sequence stars, which are thousands of times cooler and dimmer than Alnilam. Stars outside the main sequence include a dim, hot white dwarf (Sirius B), three bright but cool giants (Beta Corvi, Arcturus and Menkar) and three supergiants (Alpha Aquarii, Alpha Arae and Betelgeuse).

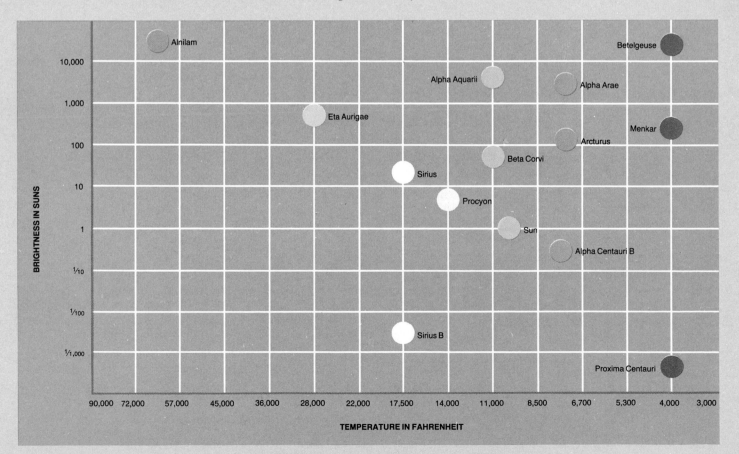

Alpha Aquarii 95,000,000 miles

Menkar 48,000,000 miles

Alnilam 27,000,000 miles

Arcturus 17,000,000 miles

Beta Corvi 9,500,000 miles

Eta Aurigae 3,000,000 miles

Sirius 1,700,000 miles

Procyon 1,500,000 miles

Sun 864,000 miles

Alpha Centauri B 648,000 miles
Proxima Centauri 216,000 miles
Sirius B 32,000 miles

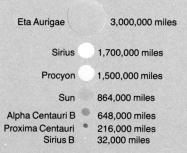

TEMPERATURE AND COLOR

The basic color of a star, visible to the naked eye
if the star is large enough or close enough,
is an indication of its temperature. The stars at
left are identified by the color swatches and
temperature ranges below.

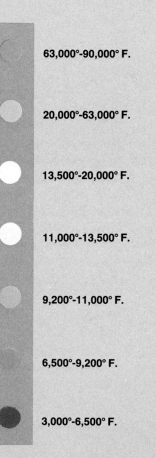

63,000°-90,000° F.

20,000°-63,000° F.

13,500°-20,000° F.

11,000°-13,500° F.

9,200°-11,000° F.

6,500°-9,200° F.

3,000°-6,500° F.

SIZE AND COLOR

Stars display apparent inconsistencies in the relationship of their size and color. The smallest
star, at bottom, is a white dwarf, Sirius B.
Above it are six main sequence stars ranging
in color from red (cool) to blue (hot). Of the
top seven stars, the largest three are supergiants, and three others are giants. The
exception is that blue supergiant Alnilam,
which is much smaller than the red giant Menkar but is classified as a supergiant because it
is brighter and has more hydrogen fuel to burn.

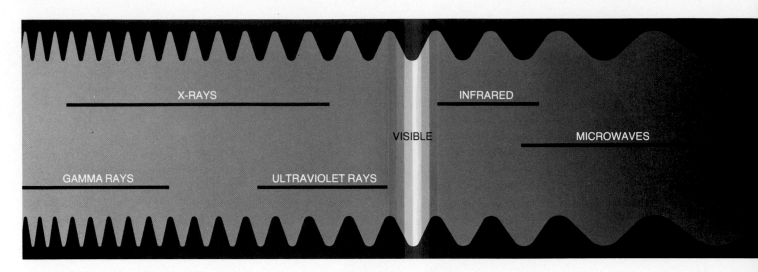

GAMMA RAYS | X-RAYS | ULTRAVIOLET RAYS | VISIBLE | INFRARED | MICROWAVES

first time the existence of an extraterrestrial magnetic field. This feat, according to Robert Woodward Simpson, the physicist president of the Carnegie Institution of Washington, "is surely the greatest advance since Galileo's discovery of those blemishes on the sun."

As the spectrograph provided the key to the sun's chemistry, that knowledge in turn led eventually to resolving the central question of solar physics: How does the sun actually produce its enormous outpouring of energy?

It was a question that taxed some of the greatest minds of late-19th Century science. The earlier and quite reasonable assumption that the sun's heat came from chemical combustion — simple burning, in other words — could no longer hold. William Thomson (later Lord Kelvin), the noted Scottish physicist and mathematician who originated the Kelvin temperature scale, calculated that if the sun were made of coal it would burn out in less than 5,000 years — a preposterously short period of time, given the overwhelming evidence from the fossil record that life had existed on Earth for millions of years. The developing knowledge of solar chemistry cast further doubt on combustion as the sun's energy source, for other calculations showed that a sun composed of hydrogen and oxygen would consume its fuel in only 1,500 years.

If combustion failed to do the trick, maybe gravity would. Newtonian physics declared that when one object falls into another, its energy of motion is not lost but transferred — much of it into heat from the force of impact. One seriously debated theory proposed that the sun might be heated by meteoroids and other cosmic debris falling into it. Unfortunately, there did not seem to be nearly enough meteoroidal material passing through the solar system to supply this force.

Meteoroids, however, were not the only potential source of gravitational energy. Perhaps the outer regions of the sun itself, by collapsing inward under the tremendous forces of solar gravity, might transfer their energy of motion into heat. This theory of gravitational collapse was proposed by the brilliant German physicist Hermann von Helmholtz in 1854. Quite correctly, Helmholtz assumed that the sun had formed from the condensation of matter "once diffused in cosmical space." He calculated that the energy resulting from the collapse of so great a quantity of matter would heat the sun to a temperature of 28 million degrees Fahrenheit. Some of this heat would have radiated away during the early eons of condensation, but

RADIO WAVES

Energy emitted by the sun spans the electromagnetic spectrum from short-wavelength gamma rays to long-wavelength radio waves. The wavelengths emitted by a particular area of the sun are an indication of its temperature. Shorter wavelengths have more energy and are therefore hotter. Only a relatively small percentage of long-wavelength rays, starting with those in the visible range, ever reaches Earth; the atmosphere blocks out the rest.

enough would have remained for the sun to radiate at its present level for at least the past 22 million years. Helmholtz estimated that the sun would keep shining at its current intensity for an additional 17 million years as its diameter contracted at the imperceptible rate of 300 feet a year.

The Helmholtz contraction theory gained wide acceptance and dominated scientific thinking into the 20th Century. Yet there were nagging problems with it. The theory accounted for a sun that was millions of years old. But the discovery of radioactivity by the French physicists Henri Becquerel and Pierre and Marie Curie in the late 1890s pushed back the estimated age of Earth and the solar system to several billions of years. It allowed the first accurate dating of fossil formations based on known rates of radioactive decay, showing that life on Earth was at least 600 million years old — more than 27 times older than the contraction theory allowed.

The discovery of radioactivity ushered in the age of atomic physics. This new science, even as it undermined the Helmholtz idea of a contracting sun, pointed the way toward a more convincing explanation for the mechanism of solar energy.

That mechanism is nuclear fusion, and its discovery followed in the wake of 20th Century developments regarding the equivalence of mass and energy and the structure of the atom. In the first breakthrough, coming in 1905, a 26-year-old Swiss patent examiner named Albert Einstein gave the world his momentous equation $E = mc^2$: Energy equals mass times the speed of light squared. In effect, Einstein was saying that a very small amount of mass contained a potentially enormous amount of energy — if only some way could be found to liberate it.

In the early 1920s, British physicist Sir Arthur Eddington determined that the temperature at the center of the sun must be millions of degrees to counter the contracting force of gravity. Such high temperatures were essential to the theory that Cornell University physicist Hans Bethe worked out in 1938. He proposed a way that hydrogen nuclei in the core of the sun could fuse to become helium and release energy.

Of course, physicists cannot directly observe what takes place at the core of the sun, any more than they can see the single proton that forms the nucleus of a hydrogen atom. Instead they rely on a rigid logic of mathematics, backed up by laboratory experiments, to construct theoretical models of the unseen physical world. In Bethe's model of nuclear physics at the solar core, the process begins with two protons fusing together to form an atom of "heavy" hydrogen, or deuterium, whose nucleus consists of both a proton and a neutron. Next, the deuterium nucleus collides with another proton to form a nucleus of "light" helium (two protons and one neutron). Finally, two nuclei of light helium fuse to form a nucleus of ordinary helium (two protons and two neutrons).

In each step of this process, some fraction of mass is lost — that is, converted to energy according to the equation $E = mc^2$. Because a proton and a neutron have the same mass, one might assume that a helium nucleus with its two protons and two neutrons would have four times the mass of the single proton in a hydrogen nucleus. But in fact, as Einstein's equation predicts and laboratory experiments confirm, a helium nucleus is just a bit less massive than it seemingly ought to be. The difference is small — .7 per cent, or 7 parts in 1,000. It is this missing mass that has been converted into energy in the fusion reaction.

The sun releases enormous quantities of energy into space. Calculations based on Bethe's fusion model show that its energy production each second is enough to supply the electrical needs of the United States for 50 million years. Expressed another way, the sun's outpouring is of the magnitude of 100 billion hydrogen bombs exploding every second.

Scientists had long since calculated that the sun's mass is approximately 2,190,000,000,000,000,000,000,000,000 tons, or the equivalent of 333,000 Earths. Using this stupendous figure and the general properties of nuclear fusion and thermodynamics, they developed a model of the sun's interior and the elaborate mechanism by which its energy gets to the surface. The massive outer layers of the sun press inward on the core with more than 200 billion times the atmospheric pressure on Earth, squeezing its hydrogen to a density more than 160 times greater than water, generating temperatures of 27 million degrees Fahrenheit — high enough to ignite and sustain the fusion of hydrogen into helium.

Under such intense pressures, the sun would quickly collapse into itself were it not for its nuclear furnace, which produces a counterpressure that keeps the solar interior in exquisite balance. If, for instance, the fusion reactions slow, the pressure on the core increases. This liberates gravitational energy, producing heat — Helmholtz was at least partly right — which causes the reactions to intensify and bring everything back in balance. If the core produces too much heat, it expands outward and cools, and the reactions slow. The sun, in other words, has a self-adjusting thermostat.

As energy is released from the fusion in the solar core, another incredible process begins: the struggle of that energy to reach the surface. At first, the energy radiates upward in the form of invisible gamma rays and X-rays that travel only a short distance before being absorbed by the dense hydrogen gas in the vast solar interior. Each time a hydrogen atom absorbs an X-ray, it emits another to rise a little farther. Eventually, because of cooler temperatures closer to the surface, the energy being absorbed and reemitted changes in wavelength from X-rays to ultraviolet radiation and, as it nears the surface, to visible light. In the outer 30 per cent of the sun's radius, much of the energy is transported upward by convection, the same process that causes bubbles to rise in a pan of boiling water.

If an X-ray produced in the solar core were somehow able to reach the surface unhindered by any part of this process, it would make the journey in 2⅓ seconds. In fact, the upward path of radiation — from the first fusion reaction in the interior to the release of light at the surface — takes about 10 million years.

With the rapid advance of scientific knowledge, the sun becomes an ever more fascinating star. Scientists have learned, for instance, that the 11-year sunspot cycle first noted by Schwabe is just half of a less visible but more fundamental 22-year magnetic cycle, during which the sun gradually reverses its magnetic polarity. Thus, during one sunspot cycle, the sun's northern hemisphere has a positive polarity; during the following cycle, the positive polarity is in the southern hemisphere. This magnetic flip-flop is thought to have its origins in the solar convection zone, where interactions of up-and-down convection eddies and the sun's rotation stretch the solar magnetic field and gradually change its orientation from one direction to the other.

This 22-year magnetic cycle has been linked, through evidence in tree

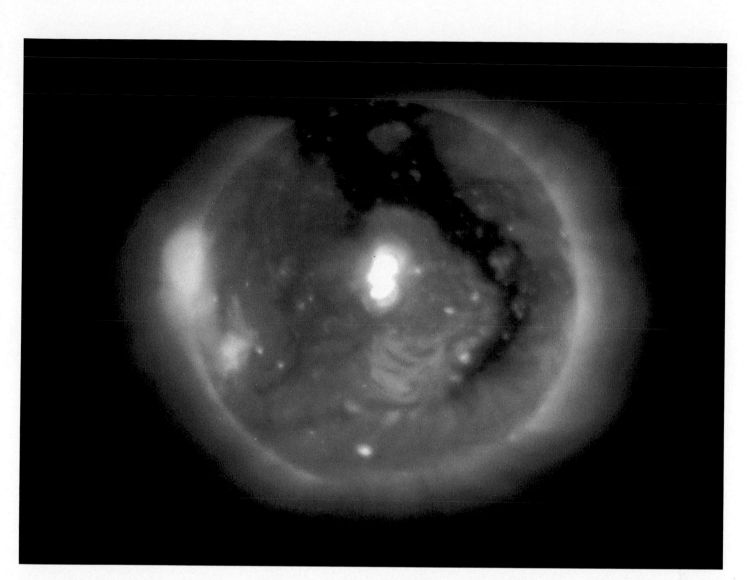

An X-ray photograph taken from the Skylab space station, in orbit outside Earth's atmosphere, reveals a dark gap in the sun's corona. Such gaps, called coronal holes, are breaches in the magnetic field of the sun. Through these gaps, the solar wind spews its stream of atomic particles into space.

rings, to periods of drought on Earth. The connection reinforces the old idea of a tie between changes in solar activity and swings in earthly climate. The absence of sunspots between 1645 and 1715 and the concurrent glacial period in Northern Europe is perhaps the best-known example of this, but the American astronomer John A. Eddy has traced similar events—18 in all—as far back as 5000 B.C. Some scientists are convinced that a cause-and-effect link exists, although no one has yet satisfactorily explained the mechanism that might produce it. Princeton physicist Robert H. Dicke has proposed that sunspots, flares and other observed phenomena at the solar surface are merely manifestations of something deeper—an internal clock that regulates the sun's 22-year cycle. But the nature of this clock, if it exists at all, remains a mystery.

This and other questions about the sun cannot be answered until new data and deeper theories are developed. Yet even if the future produces satisfactory answers, scientists and laymen alike will still gaze in wonder at this central force in the solar system. The ancients had good reason to worship the sun. It nourishes all life on the only planet where life is known to exist. Under its wondrous light, human beings bask in warmth and pale to insignificance. The planets, asteroids and comets dance in stately orbit, locked in the sun's gravitational embrace. Ω

THE SOLAR FURNACE AT WORK

Based on recent analyses of solar activity, scientists have built up a picture of the sun as a series of concentric layers that interact continuously, churned by enormous heat and immensely powerful magnetic fields. At the core, nuclear fusion generates solar energy and sends it percolating up through the radiation zone, convection zone, photosphere and chromosphere. The outermost layer is the corona, where solar gases form a burning halo and solar particles are spewed into space to enclose the planetary system in a magnetic envelope.

The invisible lines of magnetic force come from within the sun and run up through the layers, writhing, knotting and twisting in constant turmoil. These force lines are responsible for the sun's most dramatic features: powerful flares and towering prominences.

CORONA

POLE

THE CONVECTION ZONE
Before reaching the surface, energy from the radiation zone enters a layer of cooler, more opaque gas. Here energy is moved upward by convection cells: As the hot gas rises and expands, it sets up turbulent convection currents, which carry energy toward the surface in a violent, bubbling motion, accompanied by rumbling sound waves.

PROMINENCE

THE RADIATION ZONE
As energy from the core percolates upward, it passes into the radiation zone, thickest of the sun's layers, through which the energy particles move by a process of absorption and reemission. In this zone, temperature and pressure decrease, and changes in wavelength transform gamma rays into X-rays, thence into ultraviolet light and visible light.

RADIATION ZONE

THE SOLAR CORE
The core of the sun is subjected to 200 billion times the pressure on Earth's surface and is heated to 27 million degrees Fahrenheit. In this nuclear furnace, 700 million tons of hydrogen atoms per second are fused together to form helium. In the process, five million tons of matter are converted into energy.

CORE

THE CORONA

The corona, the sun's fiery outer atmosphere, is defined by magnetic loops that hold the solar gases within 240,000 miles of the surface. Although the particles here have been spectrographically analyzed as being tremendously hot — as much as five million degrees Fahrenheit — they are widely dispersed; as a result, the overall heat of the corona is considerably lower. The solar wind flows into space through holes in the corona.

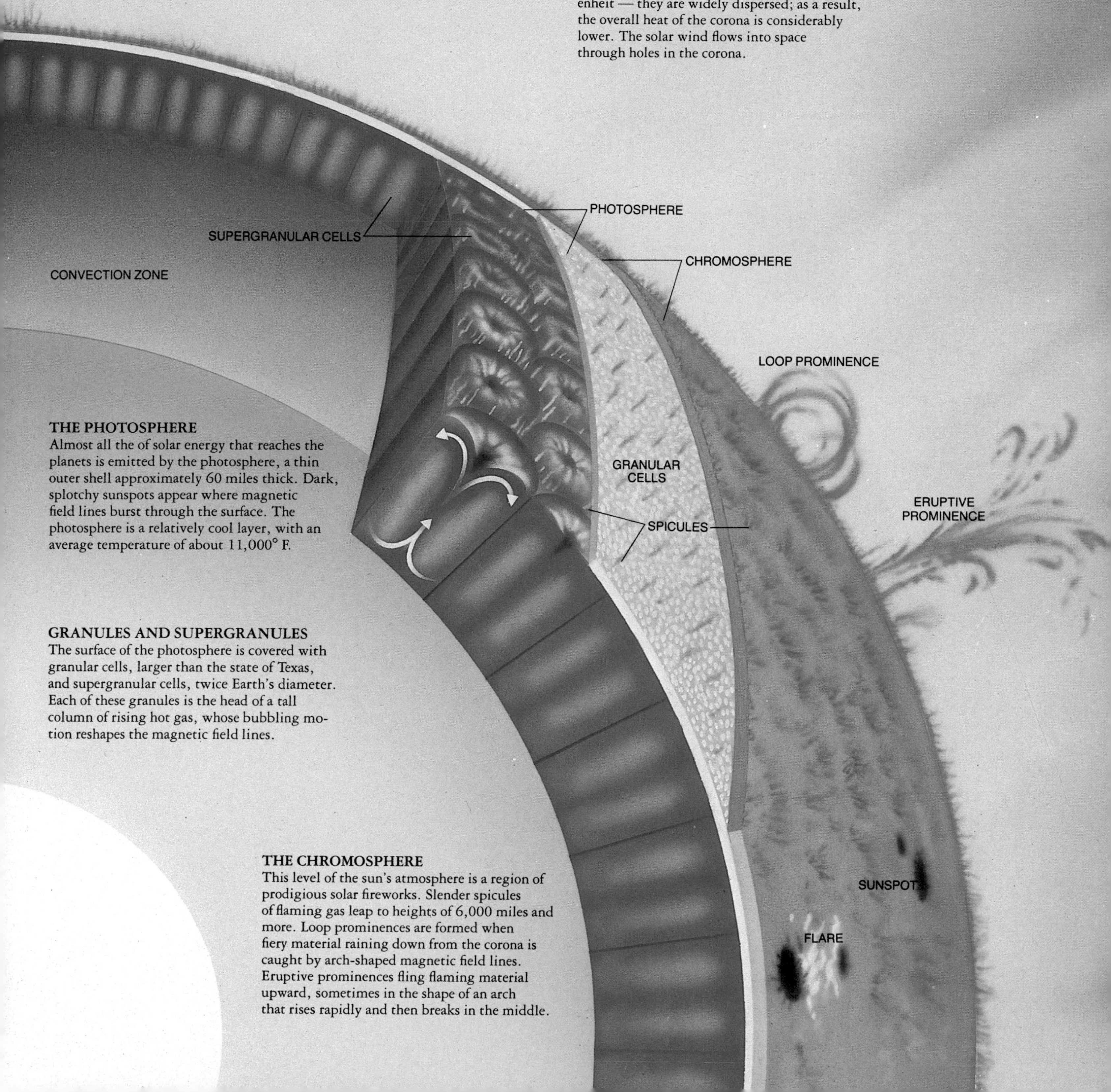

PHOTOSPHERE

SUPERGRANULAR CELLS

CHROMOSPHERE

CONVECTION ZONE

LOOP PROMINENCE

GRANULAR CELLS

THE PHOTOSPHERE

Almost all the of solar energy that reaches the planets is emitted by the photosphere, a thin outer shell approximately 60 miles thick. Dark, splotchy sunspots appear where magnetic field lines burst through the surface. The photosphere is a relatively cool layer, with an average temperature of about 11,000° F.

ERUPTIVE PROMINENCE

SPICULES

GRANULES AND SUPERGRANULES

The surface of the photosphere is covered with granular cells, larger than the state of Texas, and supergranular cells, twice Earth's diameter. Each of these granules is the head of a tall column of rising hot gas, whose bubbling motion reshapes the magnetic field lines.

SUNSPOTS

THE CHROMOSPHERE

This level of the sun's atmosphere is a region of prodigious solar fireworks. Slender spicules of flaming gas leap to heights of 6,000 miles and more. Loop prominences are formed when fiery material raining down from the corona is caught by arch-shaped magnetic field lines. Eruptive prominences fling flaming material upward, sometimes in the shape of an arch that rises rapidly and then breaks in the middle.

FLARE

Chapter 3

NEAR NEIGHBORS

"The early Earth and the early Venus were essentially alike. The present Earth and the present Venus are radically different. Why did these two bodies — twin sisters, almost — end up so dissimilar?"

The speaker, leading a recent discussion at the U.S. Geological Survey in Flagstaff, Arizona, was Harold Masursky, a senior scientist in the Survey's Branch of Astrogeology and a member of the NASA team that selected landing sites for the Apollo moon missions. Masursky went on to answer his own question and to explain some of the key differences between Earth and the four other major bodies of the inner solar system — Mercury, Venus, Earth's moon and Mars.

Of course the early Venus and Earth did have consequential differences, among them distance from the sun, and they still show significant similarities. But for all five bodies of the inner solar system, the general historic trend has clearly been one of divergent evolution. All five were formed of the same rocky materials, at the same time about 4.6 billion years ago. All five came into being close enough together — in orbits less than 106 million miles apart — to have suffered the same long bombardment of space debris. And yet, as space-age photographs and data show, the five turned out to be distinctively dissimilar in many aspects, such as atmosphere, temperature, water content, chemical composition and internal structure.

For all that has been learned about the inner planets in the past two decades, the steps by which their evolutions diverged are imperfectly understood. Still, systematic comparisons of the sort Masursky was making have proved informative in themselves. Similarities between terrain features on the planets have enabled scientists to draw sound conclusions about the type, order and duration of various geological processes. Inferences drawn from new data about other planets have helped fill in gaps in their knowledge of Earth's distant past. Indeed comparative planetology — the study of the similarities and differences between the planets — may well shed light on the future of planet Earth and the solar system.

Earth is usually studied parochially, as a series of vivid and varied parts. Viewed as just another planet, it is fully as remarkable. A smallish body, it is now unique in that three fourths of its surface is covered with oceans and seas; though both Venus and Mars once may have had quantities of fluid water (as distinct from water vapor), they no longer do. Earth's interior is highly differentiated. According to analyses of seismic waves, the crust, or lithosphere, is about three miles thick under the oceans and about 20 miles thick under the continents. The crust floats on the mantle, a region of hot,

Rivers of lava remodel the Hawaiian landscape during a violent eruption on the flank of Kilauea volcano in early 1984. Scientists believe that such volcanic activity also shaped the other three planets of the inner solar system, as well as Earth's moon.

dense rock that extends 1,800 miles into the interior. Beneath the mantle lies a core of molten iron alloy, which in turn surrounds a solid core of similar material. Circulation within the molten core produces Earth's strong magnetic field.

Oceans are only one of the unique features that distinguish Earth from the other inner planets. Another is a system of enormous, rigid plates that float on the hot plastic rock of the mantle and that are kept in motion by convection currents inside the mantle. Under the oceans, the plates move apart slowly — .39 to 5.9 inches a year — on either side of the globe-girdling Mid-Ocean Ridge, where molten material from the interior wells up to form new crust. Volcanic land platforms such as the Hawaiian Islands are also built up from the ocean floor as the plates inch past "hot spots" — places in the crust where lava oozes up or erupts explosively. On land, small amounts of new crust are formed along deep chasms such as Africa's Rift Valley, the great gash that runs from the lower Red Sea to Mozambique. Where plates collide on land, long ranges of folded mountains are formed — a terrain feature found on no other planet.

As the new crust spreads the ocean floors, the displaced far edges of old crust plunge back into the mantle in long, deep trenches known as subduction zones. The old crust is melted, recirculated by convection and extruded again. As an astonishing result, the ocean floors have been found, in a series of samples, to date back no more than 200 million years. And the constant renewal of Earth's crust maintains a state of dynamic equilibrium with the forces of erosion.

Yet another unique earthly feature is the form of matter known as life, and it is closely associated with the moving plates. The theory of plate tectonics, which was developed in the late 1960s, subsequently led to the idea that the system of circulating plates made the oceans conducive to the evolution of life-forms. While the basic connection is now unchallenged, several scientists have recently proposed a variation that is winning widespread acceptance. The theorists argue that the existence of primitive life-forms made it possible for plate tectonics to become a self-sustaining system.

As the advocates point out, Earth about 3.5 billion years ago had a thick atmosphere that shielded the shallow oceans from the sun's lethal ultraviolet rays. However, the atmosphere was inimical to life because of its heavy concentration of noxious gases, especially carbon dioxide. But primitive marine plants evolved that tolerated carbon dioxide and later breathed it, absorbing the carbon content and exhaling oxygen. As life-forms proliferated and grew complex in the improving atmosphere, the carbon dioxide they absorbed was incorporated in their shells and skeletons. "In other words," explained Harold Masursky during his planetary comparisons, "animal organisms made plate tectonics. They pulled the CO_2 out of the atmosphere and used it to make calcium carbonate and precipitated it in beds of limestone."

Had the atmosphere remained heavy in carbon dioxide, it would have prevented much of the sun's energy from dissipating, and Earth would have grown hotter and hotter. In time the oceans would have evaporated. But with more and more carbon dioxide safely locked up in layers of limestone, the oceans survived and cooled the lava flows that oozed out of the Mid-Ocean Ridge. Instead of remaining warm and buoyant, the cooled lava grew

Deformed by the ravages of geologic time, this many-layered sedimentary rock from southwestern Greenland dates back 3.8 billion years, making it the oldest such specimen ever found on Earth. Analysis of the carbon atoms trapped in the ancient formation suggests that rudimentary life processes may have been at work even earlier in Earth's history.

These two-billion-year-old fossils are the oldest examples of microscopic blue-green algae, which scientists agree are the earliest form of plant life capable of photosynthesis. Blue-green algae released huge quantities of oxygen that eventually made Earth's atmosphere hospitable to more complex forms of life.

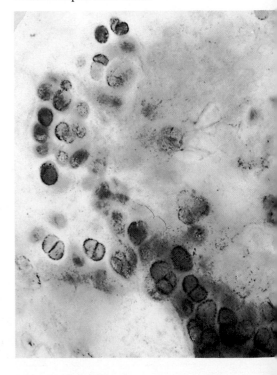

dense enough to plunge into the subduction zones, there to be remelted and reused. No one knows exactly when plate tectonics became an integrated circulation system, but it has been remodeling Earth for at least 500 million years, and possibly for much longer.

Earth's system of plate tectonics is conspicuously absent on the planet's moon. But the moon has been powerfully shaped by a force of little lasting physical effect on Earth: the impacting chunks of space debris. Friction in Earth's atmosphere burned up many of the smaller hurtling bodies, and with rare exceptions, the craters left by the larger ones have been eroded away or covered by seas, sediment and vegetation. But the moon, having no atmosphere to shield it, was heavily battered for at least 800 million years and was thereafter bombarded with slowly diminishing frequency.

Since the moon has no weather to erode its surface, countless impact craters survived for billions of years, altered only by later impacting bodies or lava flows. All this, and the splashlike rays of lunar material ejected by the crashing bodies, was obvious historical evidence, awaiting a time when telescopes were powerful enough to read the record and piece together the order of events.

By the turn of the 20th Century, the telescopes were in place. And with the dawn of the space age six decades later, scientists were able to send cameras and test instruments to scrutinize the lunar museum at close hand. Unmanned space vehicles in lunar orbit gathered much valuable information about the moon, photographing and mapping parts of the surface in greater detail than certain remote areas on Earth. Other unmanned vehicles landed on the moon and conducted simple chemical tests of the surface, revealing that moon rocks are similar in composition to Earth's basalt, a hard stone of volcanic origin. Finally, starting in July 1969 and continuing until December 1972, the Apollo program put a total of 12 astronauts on the moon and returned them safely to Earth with still more scientific data, as well as actual samples of lunar materials.

In all, six Apollo lunar landing missions brought back 842 pounds of lunar rocks and soil, taken from six locations. Three unmanned Soviet spacecraft also landed and returned with small samples taken from the eastern edge of the moon, an area not visited by the Americans. Instruments emplaced by the astronauts operated for as long as eight years, measuring moonquakes, meteorite impacts, magnetism and heat flow from the interior. Photographs and orbital data, as well as the moon rocks and information from surface instruments, have been analyzed in detail, producing a relatively complete picture of the moon's composition.

The lunar surface is covered by a layer of fine powder and broken rocky material three feet to 65 feet deep. This layer, known to geologists as regolith and popularly called lunar soil, was churned up by impacting bodies of all sizes; much of the soil itself was laid down by a steady rain of microscopic dust particles. The regolith is composed entirely of elements known on Earth, but in significantly different proportions. Lunar matter tends to be rich in such high-melting-point elements as calcium, aluminum and titanium and is poor in low-melting-point elements such as sodium and potassium. This has led scientists to conclude that the low-melting-point elements were boiled off during an early epoch when the moon was apparently much hotter than Earth.

A cyclonic storm 404 miles in diameter swirls across the Pacific Ocean in this photograph taken by an orbiting spacecraft from 120 miles aloft. Earth's

expansive seas and its varied land masses give the planet the most unpredictable weather in the solar system.

All of the lunar rocks are of two main types: igneous rock, originally molten lava, and metamorphic rock, formed when the heat and pressure of impacting bodies melted and consolidated various rocks. No trace has been found of the type of sedimentary rock that is formed by the compression of deposits at the bottom of bodies of water. The lunar rocks show no sign of water erosion. They contain no water in combination with minerals, and since the moon has no atmosphere to hold moisture, the rocks have not been weakened or altered by chemical erosion. Apparently the moon has always been waterless.

The interior of the moon is quiet. While seismometers on Earth record hundreds of thousands of tremors a year, those emplaced on the moon have detected only about 3,000 tremors annually, most of them generating little more energy than an exploding firecracker. Since many of these quakes occur at the same time each month, scientists postulate that they are caused by tidal strains as the moon swings in its orbit around Earth. All in all, the moon produces only one ten-billionth of the seismic energy released by Earth. It is, in effect, a dead body.

The interior of the moon is divided into concentric layers that resemble Earth's in structure but that differ markedly in dimension. There is an outer crust about 37 miles thick. Beneath it is a thick mantle of denser rock that extends to a level about 500 miles deep. If there is a molten metallic core, it is so small that the Apollo instruments failed to detect it.

Most of the moon rocks are extremely old. Dark lavas from the broad lunar basins have been subjected to the latest radioactive-isotope dating techniques and found to be from 3.1 to 3.8 billion years old—as old as the oldest known rocks on Earth. Rocks from the light-colored lunar highlands are even older, ranging in age from 4 to 4.3 billion years. And some tiny green rock fragments that were brought back by the final Apollo mission have been dated back 4.6 billion years to the very formation of the solar system.

In the light of all these findings, the moon's history is quite clear. The heaviest bombardment of planetary debris lasted until 3.8 billion years ago. In this period, hurtling bodies the size of Rhode Island or Delaware blasted out the huge basins. From about 3.8 to 3.1 billion years ago, great flows of lava filled the basins and hardened; in the mistaken belief that these features were large bodies of water, Classical astronomers called them maria, the plural of the Latin word *mare,* or sea. Since the lava hardened, its surface has been cratered occasionally by small- to medium-sized chunks of debris. Scientists have detected some evidence of recent small-scale volcanism in the form of younger rock fragments, but these instances are insignificant compared with the events that shaped the early moon. For all practical purposes, the history of the moon was completed three billion years ago.

All of the probings of the moon have so far failed to settle the old question of its origin. There remain three conflicting theories, each with attractive advantages and unresolved difficulties. The oldest view, and for a long time the most popular one, is that the moon was once a part of Earth, and that it somehow broke away and was hurled into orbit. This concept, known as the escape theory, was first proposed more than a century ago by mathematician George Howard Darwin, the second son of the great biologist Charles Darwin. Chemical analysis of lunar materials has turned up nothing to eliminate this possibility. On the other

Three lunar rocks, brought back from the moon by Apollo astronauts, are shown here in approximate size. They are, from bottom: a 4.6-billion-year-old troctolite that may have solidified when the moon itself was formed; a chunk of breccia, formed when the heat of impacting space debris fused a number of rocks together; and a piece of porous basalt, which rose from the lunar interior as molten lava.

hand, scientists cannot devise a convincing explanation of how the moon and Earth were separated, or how the moon thereupon settled into its orbit around Earth.

Another theory holds that the moon first appeared somewhere else in the solar system — or perhaps even beyond the bounds of the solar system — and was later captured by Earth's gravity and held in orbit. But the capture theory has its drawbacks, too. It seems highly unlikely that a moon-sized object plummeting toward Earth would be seized by Earth's gravity rather than crashing into the planet or simply rushing past it.

The third theory is that both Earth and moon were formed in their current relative positions when the solar system was created out of its matrix cloud of swirling cosmic dust. But this idea does not explain how or why two planet-sized bodies took shape so close together — only 239,000 miles apart.

In sum, the three theories appear to be evenly matched in their respective strengths and weaknesses. "We will need more data and perhaps some new theories," NASA's Bevan French wrote, "before the origin of the moon is settled."

The pockmarked face of the moon was a familiar image when, in March 1974, an unmanned American spacecraft began sending back what looked like still more pictures of the lunar terrain. But the new photographs — a total of 2,800 arrived — were really views of Mercury. The spacecraft, *Mariner 10,* had managed to reach the innermost planet by using the gravitational pull and the orbital motion of Venus as a slingshot, passing within 431 miles of Mercury. This previously untried technique had made the mission possible by eliminating the high cost of a rocket powerful enough to launch a spacecraft on a direct flight.

Mariner 10's pictures of Mercury presented the clearest views of the planet that had ever been seen. Although Mercury is one of the brightest bodies in the sky, it rarely appears above the horizon when the sky is sufficiently dark to serve as a good background. It can only be observed when it rises before the sunrise or sets after sunset, and then it appears only as a faint luminous blob. According to an apocryphal story, Copernicus on his deathbed murmured a regret that he had never seen the elusive planet.

Even so, early astronomers conducted diligent studies of the planet. They calculated that Mercury made its orbital journey around the sun in 88 days,

Dwarfed by a lunar boulder, astronaut-geologist Harrison Schmitt collects rock specimens during the final Apollo mission to the moon in 1972.

Schmitt's lunar rover, at right, weighed about 475 pounds on Earth but only 79 pounds on the moon with its low gravity.

and by the 19th Century they had deduced from the apparent constancy of the few Mercurian features they could distinguish that Mercury also took 88 days to rotate once on its axis, thereby keeping the same face to the sun. Thus was born the notion that half of Mercury was always boiling hot and half was always bitterly cold. In science-fiction stories about the exploration of Mercury, the intrepid space voyagers always land on the day-night line, where the conditions supposedly moderate each other. From there they would risk only short trips out onto the planet's torrid "dayside" or frigid "nightside."

Like so many other cherished misconceptions about the solar system, these were dashed by 20th Century technology. Using the latest ground-based radar techniques, astronomers discovered in the 1960s that Mercury's rotation period is in fact only 58.65 days, and that both sides of the planet receive their fair share of sunlight and darkness. Nevertheless, Mercury is hardly a world of salubrious climes: Midday surface temperatures can soar as high as 800° F., hot enough to melt zinc; lows on the dark side can plunge to nearly –300° F.

Mariner 10 studies confirmed earlier calculations that Mercury is made up primarily of fairly dense material—including, presumably, a large iron core. The density of this material accounts for the fact that, as revealed by close examination of *Mariner 10's* photographs, Mercury's craters are shallower than those on Earth's moon, and its ejecta patterns extend only half as far from the craters. In addition, photographs showed that Mercury has no continent-sized expanses of highlands or lowlands, as does Earth.

Mercury yielded many of its physical secrets to *Mariner 10,* which not only photographed the face of the planet but recorded information about its topography, temperature and magnetism. Along with moonlike vistas of impact craters, Mercury has some decidedly unmoonlike features. These are the long lines of cliffs, or scarps, some of them two miles high, that slash overland for hundreds of miles, cutting through whatever basins and craters lie in their path. Though the origin of these scarps is uncertain, most scientists believe that they were formed when the planet cooled and contracted, causing the overlying crust to settle and crack. The planet shows no signs of plate tectonics.

Though Mercury's core has cooled, there are indications that it is still in a molten state. At least, some scientists have drawn that conclusion from the finding that Mercury has a magnetic field. The field has about one sixth of the strength of Earth's, but its very existence is in stark contrast to the virtual absence of a magnetic field around Mars and Venus. Other researchers maintain that the planet's 58.65-day rotational period could not produce sufficient convection in the core to generate even a weak magnetic field. These doubters theorize that Mercury once had a field-producing molten core that has since cooled, and the current magnetic field is the result of permanent magnetization of the planet's surface rock.

Mercury's largest impact craters have been filled with lava, like the maria on Earth's moon. In addition, broad lava fields have been detected between the craters, suggesting that Mercury went through extensive volcanism during the period of heavy cratering around four billion years ago. So much lava paved the surface that scientists think its source must extend to a depth of hundreds of miles. This lends support to the thesis that Mercury has a large iron core and helps to account for the fact that Mercury has a density

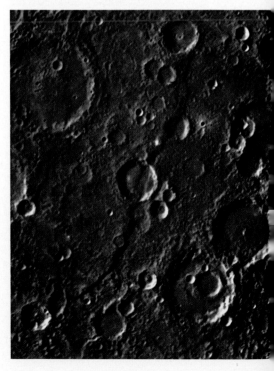

The surface of Mercury, shown in photographs taken from the *Mariner 10* spacecraft, closely resembles the heavily cratered landscape of Earth's moon. The slash in the picture below is a scarp about 300 miles long and two miles high; it was probably formed by crustal fracturing and faulting. At right, the planet appears as a sunlit crescent in a photomosaic taken by the departing spacecraft from a distance of 130,000 miles.

comparable to Earth's. However, the *Mariner 10* pictures show only one half of the planet, and many questions about Mercury will not be resolved until high-resolution photographs are taken of the other half.

While slingshooting on the way to Mercury, *Mariner 10* added considerably to the body of knowledge about Venus, the planet so similar to Earth in size and density that it has often been called Earth's twin. Venus has been familiar throughout human history. Some early peoples knew it as the bright morning star, and to others it was the bright evening star. Because of its proximity to the sun, Classical astronomers in pretelescope days did not realize that Venus was both the morning and the evening star, depending on which side of the sun it was on at the time.

Until quite recently, the planet's surface was a mystery, hidden from sight by a thick cloud cover. The clouds inspired visions of a friendly, watery planet, wreathed in oceans like Earth's. Scientists know now that the planet is a waterless and hostile world baking in hellish heat of about 900° F. and oppressed by a thick atmosphere of carbon dioxide and sulfuric acid clouds that bears down on the surface with a pressure 90 times that of Earth's.

Because conventional telescopes cannot penetrate the dense Venusian atmosphere, much information about this torrid planet has come by way of ground-based and spacecraft radar, which beamed radar waves at the planet. Radar signals reflected from high points on Venus differ from those signals that were returned from slightly lower points, making it possible for astronomers to reconstruct the features and deduce the nature of the planet's terrain.

Using radar data, radio astronomers determined that a day on Venus — the time that it takes the planet to complete one full turn on its axis — lasts as long as 243 days on Earth. Just as odd, the Venusian year — the time that it takes the planet to complete a circuit of the sun — is only 225

The basaltic rocks and soils of Venus, shown here in a ground-level picture transmitted by the Soviet *Venera 13* spacecraft, are strikingly similar to those found on many earthly terrains. The craft, partly visible at the bottom of the photograph, lasted two hours and seven minutes before succumbing to the scorching heat and crushing pressure of the Venusian atmosphere.

Earth days long; thus, to the extent that Venus experiences seasonal changes, all the year's seasons occur within a single day. Odder still, Venus rotates on its axis in a retrograde direction, so that the sun rises in the west and sets in the east.

Detailed information about this obscure planet has come not only from American space probes but also from missions carried out by the Soviet Union. During several attempts, the Soviets' spherical research vehicles were crushed by the enormous pressure of the Venusian atmosphere; even at 14 miles above the surface, the air pressure is 18 times that of Earth's at sea level. In 1970, the Soviet spacecraft *Venera 7* ejected a packet of instruments that sent back data until it crash-landed on Venus. Later, automated Soviet space vehicles managed to make soft landings on the rocky surface. Instrument analysis showed the temperature was 860° F. and that carbon dioxide made up 94 per cent of the Venusian atmosphere — as opposed to only about .033 per cent of Earth's atmosphere. Venus had just a small amount of nitrogen — the earthly atmosphere's major constituent — and only a trace of water vapor.

In another astonishing discovery, the Soviet data showed that only about 2 or 3 per cent of the solar energy reaching the Venusian cloudtops penetrates to the surface of the planet, and that very little of this infrared radiation escapes back into space. Clearly the thermal energy that does reach the surface is trapped by the carbon dioxide atmosphere and the clouds of sulfuric acid. This has produced a greatly exaggerated form of the greenhouse effect, superheating the planet and keeping it that way.

The Soviets' first Venus probes sent back only scientific measurements. Then, in 1975 and again in 1982, camera-equipped Soviet craft managed to land and beam back the first photographs taken from the Venusian surface. Each of the four vehicles that landed transmitted a single, tantalizing view of Venus and then succumbed to the planet's great heat and pressure. The pictures revealed different landscapes, one strewn with large flat rocks, a

The dense cloud cover of Venus blots out the torrid planet's surface in this ultraviolet photograph taken by the orbiting spacecraft *Pioneer Venus* from an altitude of 40,000 miles. The dark areas indicate variations in the components of the clouds or the atmosphere above them. Scientists measured cloud movement in a series of pictures and estimated the wind speed at Venus' equator to be 220 miles an hour.

second littered with stony fragments, another showing broken layers of rock. The layers could have been laid down as volcanic sediment or spread in thin lava flows; samples analyzed chemically by spacecraft instruments indicate that some rocks are similar to the basaltic lava flows that cover Earth's ocean floors in many places.

Still more information came from Venus in 1978, when the American spacecraft *Pioneer Venus* went into orbit around the planet and sent back detailed readings from a small radar instrument that slowly, strip by vertical strip, measured the planet's topography much more accurately than Earth-based radar. Computers on Earth processed the orbiter's information and incorporated it on a color-coded topographic map of 93 per cent of the Venusian surface. The detail is not great but it is more than enough to distinguish the major topographical features of Venus.

The map shows that Venus is, for the most part, uncratered, with relief similar to Earth's. More than 60 per cent of the surface is gently rolling uplands. Less than 30 per cent is lowlands. Only 10 per cent is rugged highland terrain, but its features are tremendous. A mountainous region in the northern hemisphere boasts Maxwell Montes, an ancient volcanic peak towering 36,000 feet above the surrounding plain. Along the equator lies a large region known as Aphrodite Terra, which has a rift valley two miles deep—a feature suggesting some form of tectonics. And an area known as Beta Regio may be the largest stretch of volcanic terrain in the solar system.

The sheer size of Venus' volcanoes suggests that a plate tectonics system never evolved on the planet in spite of the likelihood of ample water. Were the volcanoes located on plates that moved across hot spots in the crust, they would have passed beyond the hot spots before growing so enormous. Instead, they seem to stand fixedly above places where volcanic

The topography of Venus stands out sharply in this computer-produced global view based on radar data from the *Pioneer Venus* spacecraft. To accentuate the planet's relief, the artificial color scheme shows lowest-lying areas in blue, moderate elevations in green, higher elevations in yellow and highest in red. The apparent holes at top and bottom are areas from which no data was received.

activity was particularly intense. Any chance that Venus had of developing a plate system must have disappeared at an early date, along with the planet's shallow seas.

Some scientists, making a dazzling leap of imagination from the connection on Earth of oceans and life, speculate that life-forms may have evolved on Venus before its seas evaporated. One of the proponents of the idea is Harold Masursky. "We think that life may have started on Venus," he said. "But the planet was a little more hostile than Earth, and early life didn't develop far enough to absorb the carbon dioxide being spouted by all those volcanoes. So then the super greenhouse effect got rid of the oceans by heating them; the hydrogen component of water escaped through the atmosphere and the oxygen component got trapped in the rocks as mineral oxides. Naturally the primitive little life-forms died off."

The scientists who speculate that life could have begun on Venus now extend the same possibility to Mars, the fascinating outpost planet of the inner solar system. Ironically, the suggestion came soon after that possibility had been dismissed with apparent finality.

Mars had been viewed by wishful thinkers as the home of an advanced civilization ever since the 1890s, when astronomer Percival Lowell discerned long channels on the Martian surface and interpreted them as canals built by purposeful, well-organized beings. While most scientists doubted that the channels were artificial waterways, some believed that parts of Mars were overgrown by seasonal vegetation. Thus the possibility of some form of life still tantalized men in scientific circles.

That hope all but died when the American spacecraft *Mariner 4* flew by Mars in 1965. The craft sent back 22 pictures, taken from a distance of about 6,100 miles, of a world seemingly devoid of all the conditions associated with life on Earth. Mars had moonlike craters, but many fewer of them than Earth's satellite. It had channels, but they seemed to be waterless, random stream beds rather than constructed canals. The whole planet seemed to be defunct — as dead geologically as Earth's moon. The scientists charged with interpreting the photographs declared: "The visible surface is extremely old. Neither a dense atmosphere nor oceans have been present on the planet since the cratered surface was formed." Four years later, two more American spacecraft flew past Mars and sent back hundreds of pictures that tended to confirm that glum statement.

And then, the view of Mars again changed dramatically. On the morning of November 13, 1971, after 161 days of interplanetary flight, the 1,030-pound spacecraft *Mariner 9* went into orbit 862 miles above Mars. The vehicle carried equipment to conduct six sets of experiments, including instruments to analyze the light reflected from Mars in both ultraviolet and infrared wavelengths.

Mariner 9 also carried television cameras designed to help map the Martian landscape. But those who waited on Earth to view the images transmitted back through space faced an unwelcome delay: A planet-wide dust storm was raging over the surface of Mars, and little could be seen. It later became clear that wind-blown dust is the major agent of erosion on Mars today. Whipped up in storms that rage at more than 100 miles an hour, the dust particles scour the Martian landscape with the abrading force of giant sandblasting machines.

Making the best of the frustrating situation, scientists delayed the television mapping mission and put the time to good use studying the storm patterns and taking other measurements to improve their understanding of the atmosphere. The craft's radio tracking instruments sent a steady flow of data on such things as the shape of Mars — flat at the poles, with a slight bulge around the middle — and the planet's temperatures, which averaged about 10° F. at the equator. The cameras took the first close-up photos of Mars's two tiny moons, Phobos and Deimos, showing them to be lumpy and irregular, possibly captured asteroids.

Finally, in late December, the great Martian dust storm began to abate, and Mars was gradually unveiled. From January 1972 until October of that year, the orbiting vehicle transmitted a lengthy series of dramatic views of the Martian surface. The pictures revealed much more detail than had been recorded by previous spacecraft. They showed enormous volcanoes, sizable impact basins, huge canyons, dry channels, icecaps on both poles and, surrounding the north pole, the largest sandy desert in the solar system. Here was a Mars no one had ever seen before.

The most conspicuous features on Mars — features that in fact had been identified by earthbound telescopes — were the caps of ice on both poles. Since Mars has an axis of rotation similar to Earth's and therefore follows a similar pattern of seasons, the northern icecap expands and the southern icecap recedes when it is winter in the northern hemisphere — and vice versa during the northern summer. Also as on Earth, the southern icecap is more extensive at its maximum than is the northern one.

It was originally assumed that the caps were composed of water ice. But in the 1960s new data made it clear that carbon dioxide is the main component of the Martian atmosphere, and scientists concluded that the residual

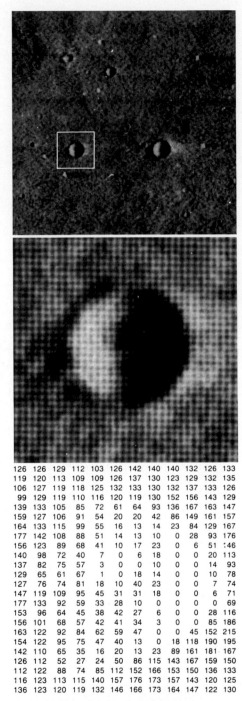

PHOTOGRAPHY BY THE NUMBERS
The visual images radioed back to Earth by far-ranging spacecraft are processed by light sensors rather than conventional photographic methods. For example, the Martian crater in the boxed area of the photograph at top is broken down by sensors into small elements known as pixels. An on-board computer translates each pixel into a number representing its level of brightness. The numbers are then transmitted to an earthly computer, which converts the pixel information back into the familiar picture.

caps, which do contain water ice, are covered seasonally with a fresh layer of crystallized carbon dioxide — dry ice.

The most remarkable thing about the poles is the terrain unveiled when the icecaps recede. Viking's close-up pictures show that the ground has few, if any, impact craters. This indicates that the terrain is geologically young — about 100 million years old. Moreover, distinct layers of sediment have been detected, and the variation in the layering suggests that climatic changes have taken place. Lively speculation began on the possible causes of climate change. "Among the most likely," wrote Michael H. Carr, leader of the Viking Orbiter Imaging Team, "are variations in the rotational and orbital motions of the planet, which cause differences in the seasonal fluctuations of solar energy and how it is distributed by latitude."

Far more spectacular than the icecaps are Mars's breathtaking volcanoes. The largest one stands alone; three others are in a northeast-southwest line. They all appear to have craters at the top. Those four volcanoes dwarf the tallest mountains on Earth. The largest, Olympus Mons (Latin for "mountain"), in a highlands area called the Tharsis Bulge, rises from a base so wide it would cover the state of Missouri, or most of England. The monster towers 78,000 feet above an arbitrarily established reference level — the Martian equivalent of Earth's sea level — and about 100,000 feet above Mars's lowest places. This means that the volcano is about two and a half times as high as Mount Everest, at 29,028 feet the tallest mountain on Earth. Olympus Mons also has more relief than the island of Hawaii, the largest similar volcanic structure on Earth, which rises more than 30,000 feet from its ocean-floor base to its summit atop Mauna Loa. Olympus Mons's total rise above the lowest places on Mars is about half again greater than the greatest relief on Earth — 65,000 feet from the bottom of the Marianas Trench, 36,198 feet below sea level, to the top of Mount Everest.

Olympus Mons owes part of its height to its location on a broad uplifted dome, about 2,500 miles in length, 1,900 miles in width and several miles above the Martian reference level. The rest is due to accumulated outpourings of lava, the same process that builds up shield volcanoes on Earth. Olympus Mons has a central crater more than 40 miles across. It closely resembles an earthly caldera — an old crater whose dome of lava collapsed of its own weight after the supply of hot lava receded. Its base is surrounded by a mighty escarpment several miles high.

A thousand miles to the southeast of Olympus Mons are the other giant volcanoes. The three, named Arsia, Pavonis and Ascraeus, are small in comparison with Olympus Mons, but nonetheless immense. Arsia Mons, the southernmost, has a smooth-floored central caldera 75 miles in diameter. Pavonis Mons, almost exactly on the Martian equator, rises 10 miles above the surrounding plains and is about 250 miles across at the base. Ascraeus Mons, the northernmost, has a many-throated central crater 35 miles across.

It may be more than coincidence that these three volcanoes straddle the Martian equator, and that Olympus Mons is not far from it. This volcanic sector has the largest known positive gravitational anomaly — an area of great mass where gravity is stronger than in other regions. So preponderant is the mass of this region that one scientist has suggested that it shifted the planet's spin axis and placed the volcanic region on the equator. As in the case of Venus' brobdingnagian volcanoes, the sheer size of these monsters is clear evidence that no system of moving plates developed on Mars; the volcanoes remained in fixed positions above hot spots and simply kept growing.

Even more astonishing than the volcanoes, perhaps, is the giant canyon system that extends more than 2,000 miles across Mars's equatorial belt and reaches depths of almost four miles. Believed to be the complex product of faulting and other processes, the Grand Canyon of Mars is so long, deep and many-branched that one of its minor tributaries would swallow up the entire immensity of Arizona's Grand Canyon.

Then there are large areas of chaotic terrain, where the broken surface was apparently modified and eroded by floods of water. Here lie many sinuous channels ranging in width from 35 miles to about 100 yards across. They looked so much like dry riverbeds that the scientists who first studied the photographic evidence were virtually convinced that the channels had been cut by water. Like rivers on Earth, Mars's channels have discontinuous marginal terraces, teardrop-shaped islands, braided stream beds and midstream bars.

Analysts discovered on close examination of the photographs that the former Martian streams apparently burst out of the ground at their fullest, then broke into thinner rivulets and finally petered out in a network of tiny trickles. The pattern is somewhat similar to that of streams in very arid deserts on Earth.

Scientists found it difficult to account for this strange phenomenon and to explain where the vanished water went. It was easy to assume that the water simply evaporated as it had on Venus; after all, Mars's present atmosphere is too thin to hold any appreciable amount of moisture. But that was not quite good enough; such vast quantities of water could never have accumulated on the surface in the first place unless Mars once had a thicker atmosphere.

Other possibilities met the eye. The higher latitudes of the Martian landscape often showed a dusting of frost in winter, and that attested to the continued presence of water in the atmosphere, though not much of it. But all the water in the icecaps and in the atmosphere would not equal the tremendous volume needed to carve Mars's many stream beds. The scientists had to look further.

An intriguing answer to the riddle eventually emerged. Dr. Duwayne Anderson of the State University of New York at Buffalo first pointed out the probability that a large part of the water that exists on Mars today is permanently frozen in the planet's substrata. Among other scientists who espouse the theory is Eugene M. Shoemaker, a geologist at the U.S. Geological Survey, a professor at the California Institute of Technology and a planner of the geologic investigations for the Apollo missions and earlier unmanned lunar missions. "I think Mars still has most of its water," Shoemaker said. "It can be found over most of the planet. Comets originally brought it in as ice in tremendous volume; then it melted and soaked in by a very complicated mechanism. Over geological time, it became distributed in various layers of sediment and between lava flows. Now it's a subsurface mass of ice miles thick, and analogous to permafrost in Alaska." It is possible that periodic regional warming by volcanic heat or heat from impacting bodies could have melted enough permafrost to cause the chaotic terrain

A rock-littered Martian landscape stretches beyond the *Viking 2* lander's radio antenna (*top*) and other equipment. The checkerboard color-calibration charts painted on the lander's side enabled scientists to assess the accuracy of color data sent back by the *Viking 2* cameras.

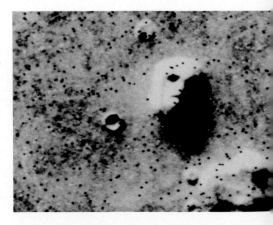

A strange rock formation on Mars, photographed from the *Viking 1* orbiter in 1976, bears a startling resemblance to a carved human face. But the fanciful notion that it was some artifact of a vanished civilization was dispelled by photogeologic interpretation, which proved it to be a shadow-dappled mesa one mile across.

to collapse and send torrents of water downslope, carving out stream beds along the way.

While American scientists were fitting together their detailed images of the Martian landscape, their counterparts in the Soviet Union produced another important bit of information about the planet. As a Soviet space vehicle passed Mars in early 1974, it radioed back data indicating that the planet had barely any magnetic field; indeed its field was scarcely stronger than that of empty interplanetary space. This information was not new; it simply confirmed evidence supplied by NASA's *Mariner 9* spacecraft, which sent back numerous magnetic readings together with more than 6,000 photographs that permitted precise mapping of the planet. But scientists could now conclude confidently that Mars does not have a molten core.

Martian horizons were widened still further by the successful Mars landing of the *Viking 1* mission in July 1976. For the first time, equipment was in place that could provide color pictures, and from ground level on Mars. Among the camera's revelations: The Martian sky is not blue, as nearly every scientist and space artist had anticipated, but pink. Apparently the

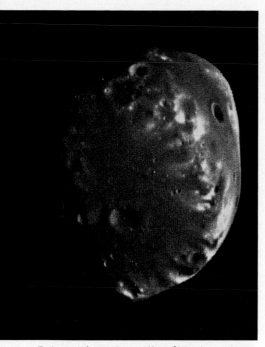

Deimos (above), the smaller of Mars's two irregularly shaped moons, measures less than 10 miles in its longest dimension. Deimos' many small craters, none more than 1.4 miles wide, are partially covered by pulverized rock, which also coats the slopes along lunar ridges.

The Martian moon Phobos, five miles longer than its lunar sister Deimos, is dominated by the 6.2-mile-wide Stickney Crater, at top, and by smaller craters in the shadows at left center and at bottom. This photomosaic also reveals some of Phobos' many long grooves, which are presumably fractures caused by the impacting body that gouged out Stickney Crater.

sky is reddened by iron oxide in soil particles raised by the wind and suspended in the lower atmosphere. The Martian surface is a deeper red.

About six weeks after the *Viking 1* lander arrived on Mars, it was joined by a sister craft, *Viking 2,* which settled down in similar desert terrain 900 miles to the north. In addition to transmitting photographs, the two landers monitored the Martian weather, tracing daily and seasonal temperature variations, which ranged from a low of $-189°$ F. to a high of $-9°$ F. But the most intriguing operation carried out by this pair of electromechanical explorers was the search for signs of Martian life-forms.

In hopes of detecting simple forms of life, the mission planners had fitted the landers with compact, sophisticated devices designed to scoop up Martian soil and subject it to three separate biological tests for the existence of organic matter. The tests would determine whether the samples showed any trace of metabolism, the vital chemical process that characterizes all life.

The results were disappointing. The Viking landers did not find concrete evidence of life at either site. However, one of the experiments produced a startling result. "When the sample was humidified," recalled project scientist Gerald A. Soffen, "it released an extraordinary burst of oxygen. This was not anticipated, and no terrestrial soils have ever done this." Such a reaction seems to indicate the presence of some kind of oxidizing substance in the Martian soil. If the soil really has an oxidizing effect, that condition would be destructive to organic substances and could explain why the Viking landers did not detect any organic material on Mars. Soffen, who once was optimistic about the chances for life on Mars, now believes that it is highly unlikely. "One doubt lingers," he wrote. "We have not visited the polar regions. I have always believed that in the search for life you must go where the water is. The permanent polar caps of Mars are frozen water and would act as a splendid 'cold finger' to trap organic molecules. Who knows what those lucky future explorers of Mars will find there?"

Even if relics of primitive life-forms are one day found on Mars, something obviously went wrong on the planet; life was a failed experiment there. If life started on Venus, it failed there too, and it undoubtedly never began on Mercury. All the more remarkable, then, that it began on Earth and survived there. The more that is learned about the other planets, the clearer it becomes that Earth was signally favored in many ways.

There are sound reasons for Earth's good fortune and its near neighbors' bad luck. Don L. Anderson, professor of geophysics and director of the Seismological Laboratory at the California Institute of Technology, proposed that Mars and Earth's moon, being smaller than Earth, cooled more rapidly after their formation and developed a solid surface layer that was too thick and buoyant to break into pieces and serve as renewable crustal plates. Anderson wrote that low atmospheric pressures, as found on small planets, and high temperatures, as found on Earth-sized Venus, worked in complicated ways to prevent the start of plate tectonics on those planets. In general, planets cool as they age, and they cannot develop moving plates if they are too old and cold, or too young and hot.

It seems that Earth was providentially placed and was vouchsafed conditions that kept it ever young and hospitable. "If the Earth were smaller, hotter, much younger, or much older," said Anderson, "conditions apparently would not be appropriate for plate tectonics." And without plate tectonics, higher life-forms might never have evolved on Earth. Ω

THE CONTORTED FACE OF MARS

June 23, 1976, was a historic day for the Viking Orbiter Imaging Team at the Jet Propulsion Laboratory in Pasadena, California. The team's 50 scientists and technicians, led by astrogeologist Michael H. Carr, received the first of more than 50,000 pictures of Mars to be sent by two Viking spacecraft in orbit around the planet. The images, which were transmitted in digital form, were reconstructed in keyed sequence. These were corrected to remove transmission errors and distortion caused by oblique camera angles. Finally the pictures were assembled in stunning topographic mosaics.

The photographs revealed — as the examples on these pages show — an awesome diversity of enormous landforms. Among the most spectacular features are three volcanic regions: the Tharsis Bulge

straddling the equator, the Elysium area in the northern hemisphere and, in the southern hemisphere, rugged terrain bordering the 1,000-mile-wide Hellas basin, the largest impact crater on Mars. From the Tharsis Bulge, a vast network of radial fractures spreads outward in all directions. To the east of the Bulge runs a prodigious system of interconnected canyons, the Valles Marineris. At its eastern end, the canyon system merges with chaotic terrain where the ground has collapsed.

Michael Carr points out that Mars has much taller mountains and much deeper canyons than Earth, and that for eons the red planet has suffered much less erosion than has Earth. "As a consequence," he says, "Mars, despite its smaller size, has considerably more relief."

Two photographs of Mars's surface show the range of images sent back by the Viking spacecraft. In the long-range picture *(below)*, taken by one of two Viking orbiters, sunlight spreads across the planet, illuminating the volcano Ascraeus Mons, a round blob at left, and the great central gash of the Valles Marineris and the frost-dusted south polar region, at right. In the ground-level photograph *(right)*, taken by the first of two Viking landers, wind-blown dust clings to rocks in a desert called Chryse Planitia. The large boulder, named "Big Joe" by the Viking Imaging Team, is three feet high.

Clouds fill the 75-mile-wide crater atop Arsia
Mons, one of Mars's four greatest volcanoes. Arsia and two of its neighbors, Pavonis Mons
and Ascraeus Mons, each cover an area 250 miles
across. They are arrayed in a line about 1,200
miles long near the crest of the enormous Tharsis
Bulge. The Bulge is more than six miles higher than Mars's mean surface level, and the volcanoes tower another 10 miles above that. Larger
still is the volcano Olympus Mons, 1,000 miles
to the northwest at the fringe of the Bulge.

The northeastern edge of the Tharsis Bulge is
marked by two comparatively modest volcanoes.
The larger of these, Ceraunius Tholus, has a
crater just 14 miles in diameter. On the flanks of
both volcanoes, channels, crater chains and
lava flows can be discerned. The fractures at left
are part of a pattern of similar lines that
extend radially from the crest of the Bulge for
thousands of miles, affecting almost half the
planet. The Bulge itself covers roughly the area
of the United States, including Alaska.

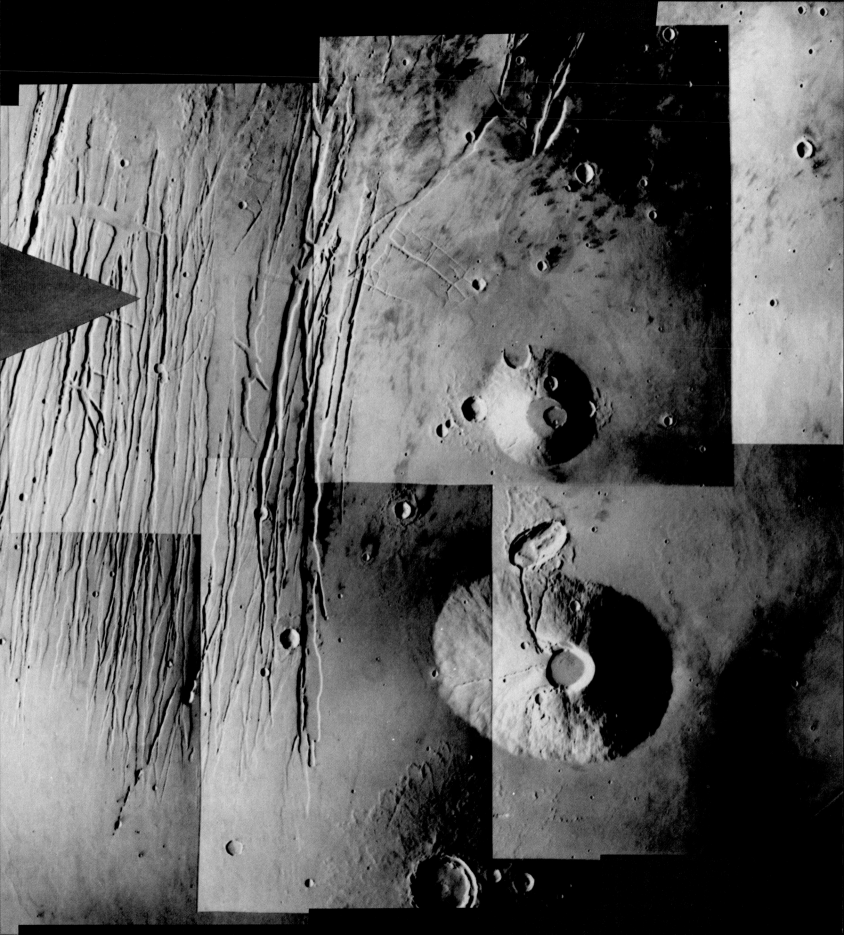

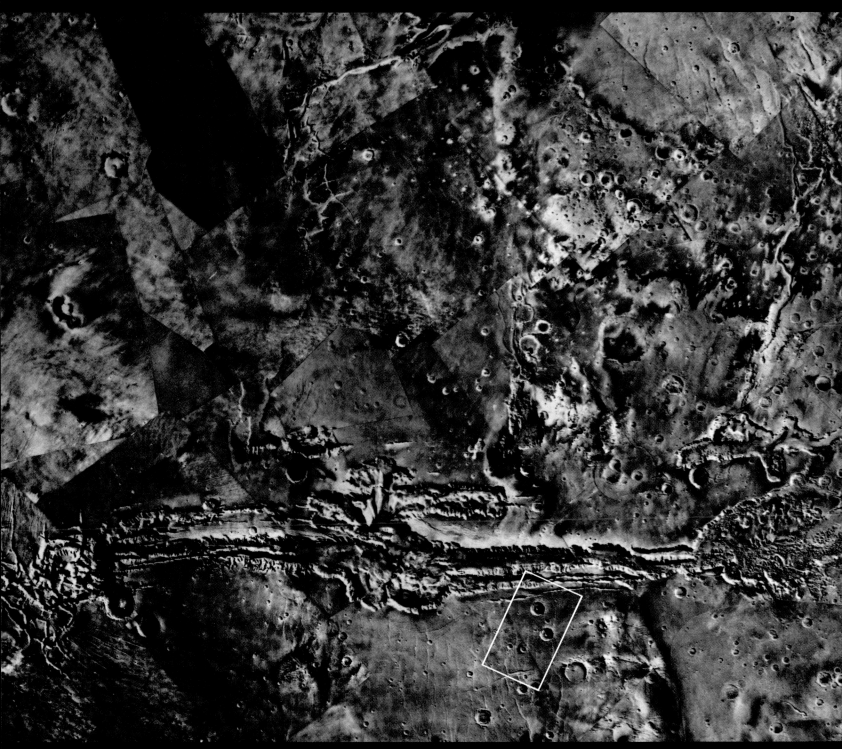

The Valles Marineris, a chain of canyons 2,500 miles long, cuts across the lower half of this mosaic. The canyon system, which reaches depths of up to four miles and measures more than 400 miles wide at its broadest point, terminates in the west in a maze of gorges at the crest of the Tharsis Bulge.

The color-enhanced view at right of part of the Valles Marineris (*box, above*) reveals the canyon's flat floor and little evidence of the sort of water erosion that created Earth's Grand Canyon. Massive faulting is thought to have formed much of the system, which was enlarged by landslides perhaps caused by ground-water seepage.

A network of intersecting lines, resembling an aerial view of fenced-off fields on Earth, marks a region of chaotic terrain. The terrain here apparently collapsed in the distant past, leaving large blocks of ground at different levels from those around them. The comparatively smooth area on the right is a lava plain.

Mars's cap of water ice at the north pole ex-
hibits a spiral pattern caused by a combination of
swirling winds and the heat of the summer
sun on south-facing slopes. The ribbon-like dark
areas are rock, exposed when the ice evaporates. Most of the water ice survives through the
northern summer, though the frozen carbon
dioxide that covers it each winter evaporates.

Yuti, an 11-mile-wide impact crater, is
ringed with a flower-like pattern of ejecta that
spewed from the crater when the meteorite
that formed it smashed into the Martian surface.
The pattern suggests that the ejecta was a
thick fluid, probably a slurry of crushed rock and
water from the subsurface permafrost that
was melted by the crashing meteorite. The cone
in the center was formed when the terrain re-
bounded from the meteorite's impact.

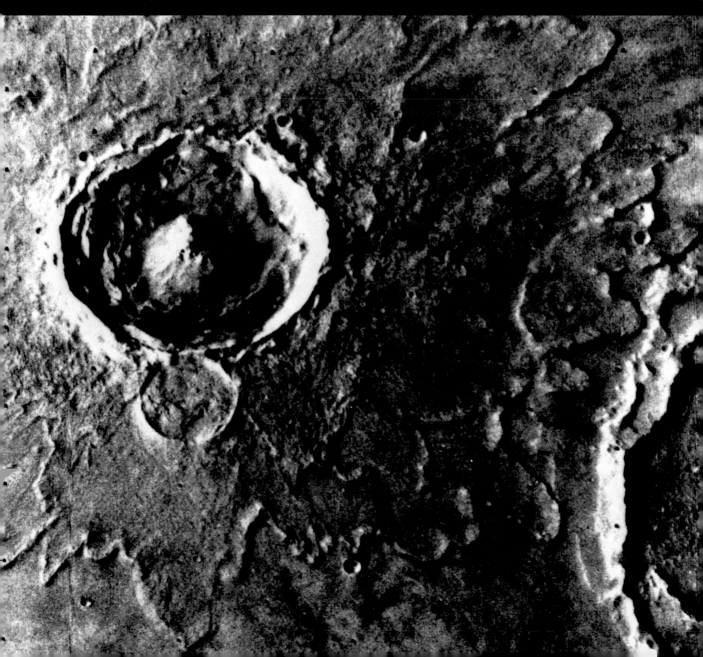

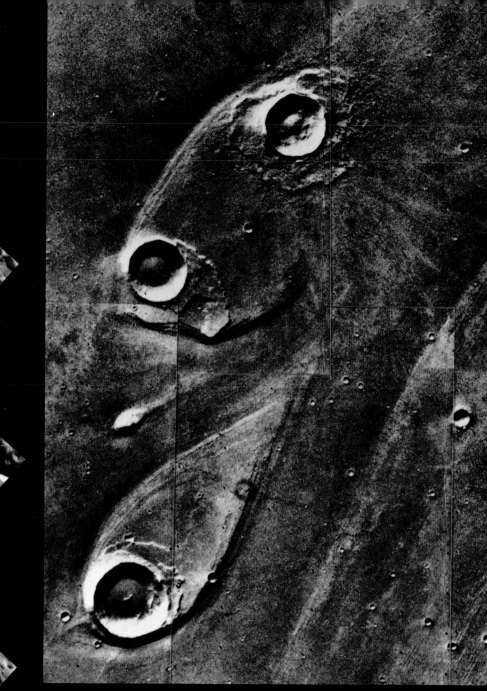

Teardrop-shaped islands *(above)*, about 25
miles long and dotted with impact craters, rise
from what seems to have been the stream bed
of a giant watercourse. There is now no liquid
water on Mars. But large volumes of water
once flowed across the surface, cutting channels
and forming islands along the way.

In an early eon, liquid water from under a
thick permafrost layer cut different types of
channels. The braided stream bed at right
carried the converging flow from its broad source
above. The network at left dispersed the water widely. Both systems resemble the flash-
flood channels in Earth's desert areas; they
are broad at their source and break into many
narrow, dwindling streams.

113

Chapter 4

GAS GIANTS AND ICEBALLS

In November 1977, following NASA's computerized flight plan, *Voyagers 1* and *2* crossed Mars's orbit, then penetrated the asteroid belt and hurtled on toward the great gaseous worlds of Jupiter and Saturn. With the sensational Martian discoveries of the Viking spacecraft fresh in their minds, scientists held out high hopes for more of the same when the Voyagers reached the two giant planets, then went on to their final objective, a rendezvous with Uranus in 1986.

Though no one realized it at the time, the Voyagers would revolutionize astronomy as thoroughly as had Galileo's telescope. Scientists were to see, through the surrogate eyes of the Voyagers' imaging cameras and data-collecting systems, worlds that a scientist later said were "beyond the sum of our collective imaginations." By the time the Voyagers finally complete their missions, they will have revealed the outer solar system to be a far different place from what it seemed to be when the little space vehicles were launched from Cape Canaveral in the late summer of 1977.

To be sure, there was a great deal left for the Voyagers to discover. Two millennia of visual study and nearly four centuries of telescopic observation had produced only basic information about Jupiter and Saturn, and the important data collected earlier by two Pioneer spacecraft came more as an appetizer than a feast. Uranus and Neptune, which were discovered in 1781 and 1846 respectively, were still little known, and even less had been learned about remote Pluto since it first appeared on Clyde Tombaugh's photographic plate in 1930. It was clear, however, that the five outer planets share certain characteristics and differ from the inner four in several significant ways.

The outer planets are more gaseous and icy than the inner planets. They are, with the exception of Pluto, much larger than the inner planets. Jupiter's diameter is more than 11 times Earth's, Saturn's more than nine times, Uranus' just over four times and Neptune's nearly four times Earth's. Pluto is a tiny body, no bigger than Earth's moon, but in keeping with its position as the farthest planet from the sun, it is probably composed very largely of ice. But for reasons well known to the scientists, the Voyagers' images of the two gas giants were bound to be somewhat disappointing: Both planets are encased in thick clouds and would never be seen. As if to make amends, the two would present a panoply of moons that were far more exciting than the scientists dared hope.

With Jupiter and Saturn leading the way with 16 and 17 known satellites respectively, the outer planets have more moons than the inner planets.

Ring systems like those of Saturn, color-enhanced here to highlight subtle differences in their composition, are a prominent characteristic of the outer planets. At least two of the other four outer planets are known to have rings, albeit less spectacular than Saturn's.

In addition, Uranus has five satellites, with indirect evidence suggesting that there may be more. Neptune, like Mars, has but two moons, and Pluto only one. Besides possessing many moons, Jupiter, Saturn and Uranus have more or less extensive systems of rings, and scientists have not ruled out the possibility that Neptune may have rings as well. These planets are ring-rich and moon-rich because they were formed so far from the sun; the sun's gravity and ultraviolet radiation, which were strong enough to destroy nascent rings and moons orbiting the inner planets, were much weaker in the outer reaches of the solar system.

Because of their gaseous, icy composition, the outer planets are much less dense than the rocky spheres of the inner solar system. Saturn has such low density that it would float in water. Nevertheless, in terms of mass, the outer planets are overwhelmingly greater than the inner ones. Together with their rings and moons, the five contain 99.9 per cent of all the planetary mass in the solar system.

Jupiter, being the closest and largest of the outer planets, has commanded avid scientific interest ever since the ancient astronomers tracked its stately progression across the skies. Actually, the Greeks and the Romans had only a vague idea of the size and distance of the giant. But they did sense that it was majestic in scale and fittingly named it after their ruling deity. They saw Jupiter as the brightest planet in the night sky; though Venus is actually brighter, the ancient astronomers saw it only in the early morning and early evening because of its proximity to the sun. Jupiter, however, shone with a steady white light, shifting its position only slowly in its 12-year orbit around the sun.

Jupiter's august standing in the planetary ranks rose even higher in 1610, when Galileo trained his new telescope on the cloud-shrouded giant and discovered to his astonishment that it was circled by four large moons. These so-called Galilean satellites were named after the lovers of the royal god in Greco-Roman mythology. They are, in order of their distance from Jupiter, Io, Europa, Ganymede and Callisto.

Saturn, too, was a familiar sight in the ancient skies; the Romans named the planet for their god of sowing, to whom they paid homage in the Saturnalia, an orgy of drinking and revelry. Saturn's 29-year orbit around the sun approximated a human life span in those days and provided the ancients with a useful natural time period. The planet's slow pace convinced astronomers that its orbit ran at or close to the outer edge of the solar system.

Galileo also discovered Saturn's rings, though he did not realize what they were. When he first glimpsed the rings, they appeared in his crude telescope as a pair of seemingly stationary satellites on either side of the planet. And, mysteriously, they vanished two years later. The mystery was solved in 1655 by the Dutch scientist Christiaan Huygens. Using a telescope of superior resolving power, Huygens saw that Galileo's two strange objects were in fact the opposite sides of a thin, flat ring nowhere touching the body of the planet. Huygens showed that the plane of Saturn's rings tilts up and down according to the planet's position relative to Earth. The rings seemed to disappear whenever they tilted edge-on to Earth, an event occurring approximately every 15 years.

Huygens also discovered that Saturn was orbited by a moon, which eventually came to be called Titan for the generation of Greek gods preceding

VOYAGER'S HIGH-TECH TOOL KIT

The two Voyager spacecraft are powered by a trio of thermoelectric generators and equipped for numerous scientific tasks with sensing and measuring devices whose names, labeled on the diagram at right, are nearly as complex as their functions. Each craft has a photopolarimeter, an ultraviolet spectrometer, an infrared interferometer spectrometer and two specialized cameras to gather radiant energy from its targets and to provide information on the chemical composition, physical form and atmospheres of the planets and their moons. Each vehicle has a magnetometer, a plasma detector, a cosmic ray detector and a low-energy charged-particle detector to collect data on magnetic fields and charged particles. The whiplike radio astronomy and plasma wave antennas of each spacecraft listen for planetary radio emissions and measure turbulence in the flow of plasma particles. Radio signals relaying all data are sent earthward from the Voyagers' 12-foot radio dishes, and interference with the beam by planetary atmospheres, ionospheres and rings also provides information about those phenomena.

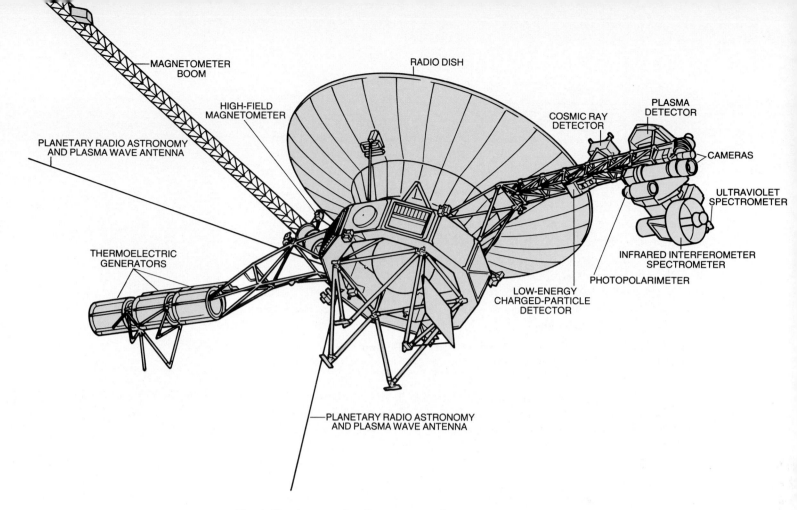

MAGNETOMETER
BOOM

RADIO DISH

PLASMA
DETECTOR

HIGH-FIELD
MAGNETOMETER

COSMIC RAY
DETECTOR

PLANETARY RADIO ASTRONOMY
AND PLASMA WAVE ANTENNA

CAMERAS

ULTRAVIOLET
SPECTROMETER

THERMOELECTRIC
GENERATORS

INFRARED INTERFEROMETER
SPECTROMETER

PHOTOPOLARIMETER

LOW-ENERGY
CHARGED-PARTICLE
DETECTOR

PLANETARY RADIO ASTRONOMY
AND PLASMA WAVE ANTENNA

Zeus (Jupiter to the Romans) and the other Olympians. Astronomer Gian Domenico Cassini, lured from Italy to serve the "Sun King," Louis XIV, as first director of the Paris Observatory, found four more satellites — Iapetus, Rhea, Dione and Tethys. These and other Saturnian moons found later were named for various deities and giants in Greek mythology.

In 1675 Cassini capped his observations of Saturn with discovery of the first sign of structure in its rings. There seemed to be a division or gap in the ring about two thirds of the way toward its outer edge. This more or less empty gap is now known as the Cassini Division. The material outside the gap became known as the A ring and the material inside it the B ring. Seven major ring sections were discerned later and identified with letters in order of their discovery.

The perplexing question of the rings' composition was debated by astronomers for the next two centuries. Cassini suggested that the rings might be swarms of small satellites, but most of his colleagues felt that they were solid bands of material. Late in the 18th Century Pierre Simon de Laplace pointed out that a solid rotating disk would be ripped apart by Saturn's gravitational forces; he postulated the existence of narrower solid bands. Finally, in the mid-19th Century, French mathematician Édouard Albert Roche and Scottish physicist James Clerk Maxwell worked out the answer. Roche calculated that tidal forces would reduce to fragments any satellite brought too near a large planet. Maxwell proved mathematically that not even narrow solid rings or fluid rings could exist, since gravitational and mechanical forces would break them apart. That left as the only reasonable hypothesis a system composed of billions of small objects that collectively give the appearance of a solid ring. The hypothesis has since been confirmed by spectrographic analysis, which later revealed that the particles are made up primarily of water ice.

Knowledge of Jupiter and Saturn grew steadily if undramatically until the 1970s; then came the quantum leap forward with the Pioneer and Voyager missions. So voluminous and detailed was the Voyagers' new information that, several years after *Voyager 2's* encounter with Saturn, scientists estimated that it would take a decade to sort out the data.

The identical Voyager spacecraft are triumphs of modern engineering. Each has the size and weight of a subcompact car and is packed with monitoring instruments, including high-resolution television cameras, spectrometers, and cosmic ray and particle detectors. Their 23-watt transmitters, whose power is equivalent to a refrigerator light, are sufficient to send 115,200 bits of data a second across the nearly 400 million miles between Jupiter and Earth—a distance so great that transmissions at the speed of light took 40 minutes to arrive.

In January 1979, the *Voyager 1* telescopes were trained on Jupiter, starting the pre-encounter "observatory" phase of the mission. Although the spacecraft was then 37 million miles from its target, the images it sent back were far more detailed than any picture taken from Earth. By the end of February 1979, with Voyager 3.7 million miles from Jupiter, the colorfully banded planet more than filled the frames of the imaging cameras. Embedded in the swirling orange and cream bands of the Jovian atmosphere, the immense, semipermanent anticyclonic storm known as the Great Red Spot grew ever larger and more spectacular. Viewers of the television monitors at the California Institute of Technology's Jet Propulsion Laboratory were hard put to describe the vast, mottled apparition looming before them. Jupiter, they said, was a "huge abstract painting," a "van Gogh" and—more mundanely—a "salad" and a "medical school anatomy slide."

On March 4, the day before *Voyager 1* was to begin the close-encounter phase, its imaging cameras made their most surprising discovery—that Jupiter has rings. Most scientists had predicted that no ring system would be found, and it was only in the interest of thoroughness that the cameras had been programed to search for rings along Jupiter's equatorial plane. And there they were: a narrow, bright outer ring and a wide, diffuse inner ring that extends all the way into the Jovian atmosphere. The system, made up of myriad minuscule particles, was extremely thin and very nearly invisible. According to astronomer David Morrison of the University of Hawaii, "The Jupiter rings are so tenuous that they would block out only a millionth of the light passing through; they are 10,000 times more transparent than the best glass." The ring system appeared in images because it was strongly backlit by the sun and the ring particles scattered the light.

Close-up photographs from Voyager showed that Jupiter is a world racked by atmospheric violence. According to measurements of the images of the Great Red Spot, its whirling turbulence covers an area 16,000 to 25,000 miles long and 9,000 miles wide and extends 15 miles above the surrounding clouds. Lesser high-pressure storms revolve like huge white ball bearings caught between jet streams roaring in opposite directions at speeds exceeding 225 miles per hour. Yet the atmosphere is not chaotic, as it had seemed at first; an orderly pattern began to emerge. The Great Red Spot apparently swallows up other storms and tends to stabilize the atmosphere. Indeed it appears that Jupiter's weather runs in tidy cycles lasting about six years. Though telescopic observation had indicated that the weather pattern had been changing steadily for the past several years, the

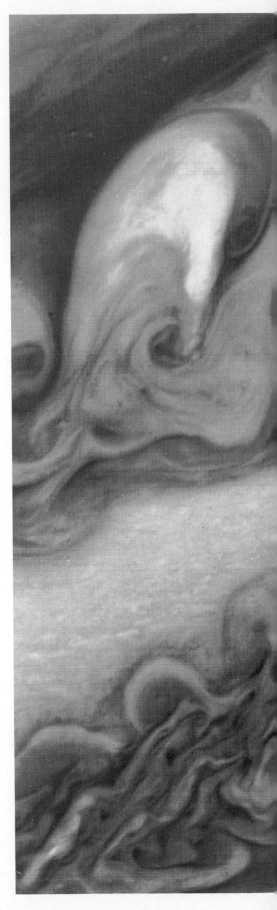

The Great Red Spot, largest of Jupiter's storm systems, is at least 16,000 miles long. It has sustained its size for hundreds of years, possibly by swallowing up lesser storms.

Voyager images seemed to be approaching those taken five and six years earlier by the Pioneer spacecraft. Whatever caused the phenomenon, it was just as though the cycle was nearing completion.

Theoretic models based on gases known to be present in the atmosphere suggest that Jupiter's massive liquid-hydrogen interior is covered by three thin tiers of clouds: a bottom layer of water ice or liquid water droplets, a middle layer of ammonium hydrosulfide (a foul-smelling compound of ammonia and hydrogen sulfide) and a top layer of ammonia-ice particles. While the upper clouds are exceptionally cold — on the order of $-240°$ F. — parts of the lower atmosphere bask in earthlike temperatures as high as $80°$ F. Measurements of the heat energy from Jupiter indicate that the planet radiates twice as much heat as it receives from the distant sun.

The composition of Jupiter — and of Saturn as well — offers an intriguing comparison with the sun. The ratio of their dominant elements — approximately nine parts hydrogen to one part helium — is roughly the same as the sun's. Unlike the sun, whose energy results from the nuclear fusion of hydrogen into helium, the radiant energy of the gas giants is mainly the leftover product of contraction from gravitational collapse during their early formation. Calculations show that Jupiter at its birth 4.6 billion years ago radiated 10 million times as much energy as it does today and was bright enough to light up space like a miniature sun. If either Jupiter or Saturn had been 10 times more massive, scientists believe, their contraction would have been sufficiently powerful to trigger and sustain nuclear fusion at their cores. Instead of one sun, the solar system would contain three; it would have evolved in an utterly different manner, and the conditions that led to life on Earth might never have occurred.

As it is, Jupiter exerts powerful magnetic forces throughout a vast surrounding region of space. The Jovian magnetosphere is so large that, if it were visible, it would appear in the sky larger than Earth's full moon. The energy in the magnetosphere consists of charged particles from three sources: the solar wind, Jupiter's ionosphere and, in greatest volume, the planet's moon Io. Atoms from the satellite's volcanoes escape into Jupiter's magnetosphere at the rate of nearly a ton per second. The particles become ionized and form a doughnut-shaped ring, or torus, around Jupiter, centered on Io's orbit.

One by one, most of the Jovian moons emerged in dramatic close-up on the imaging screens at the Jet Propulsion Laboratory. "For the planetary geologists, it's truly Christmas Eve," said Laurence Soderblom of the U.S. Geological Survey, the deputy leader of the Imaging Team. Soderblom's colleague, Bradford A. Smith of the University of Arizona, put off reporters' questions on scientific details of the Voyager data. "It may sound unprofessional," he said, "but a lot of people up in the Imaging Team area are just standing around with their mouths hanging open watching the pictures come in, and you don't like to tear yourself away to go and start looking at numbers on a printout. We will do that, but in the meantime we're just caught up in the excitement of what's going on."

The first pictures of Io, the innermost large moon in the Jovian system, were greeted with whoops of joy and amazement. Nothing like this strange orange-and-yellow sphere had ever been seen in the solar system. It was pockmarked with brown and black blemishes that seemed to defy explana-

Jupiter's recently discovered ring system, its tiny particles illuminated from behind by sunlight, stands out slender and bright in a photograph taken by *Voyager 2* from 900,000 miles.

tion. Members of the Imaging Team described its appearance as "grotesque" and "diseased" and "bizarre." To several, including Smith, it looked like a giant pizza.

The key to the enigmatic moon was supplied three days later when *Voyager 1* took a picture looking back at Io from a distance of 2.8 million miles. Linda Morabito, an optical navigation team engineer, examined the picture with her computer-controlled image display and found what appeared to be a wispy, umbrella-shaped cloud hanging above Io's outer edge. Since Io has no atmosphere, a cloud could not form, let alone be sustained. But during the next two days, scientists began to understand the curious phenomenon: Voyager had caught a volcano on Io in the act of erupting. The cloud was actually an immense volcanic plume rising more than 150 miles high and falling in a geyser-like shower. The Imaging Team examined other, computer-enhanced pictures of Io and found more volcanic plumes, a total of eight in all. As further confirmation, infrared and spectroscopic scans picked up spots of heat corresponding to the volcano sites and detected sulfur dioxide above one of them. The volcanoes ejected material at temperatures up to 710° F. into a world whose surface temperature was −230° F. It was the first time that active volcanoes had ever been found beyond Earth. Moreover, it soon became clear that Io

121

presents the most spectacular volcanic pyrotechnics of the solar system.

The material ejected by Io's volcanoes is sulfur and various volatile compounds heated to a liquid and gaseous state by a complicated process. Io, traveling an eccentric orbit, is constantly reshaped by Jupiter's tremendous gravitational pull and the gentler tug of other Galilean satellites. This heats the moon's silicate interior, which liquefies and vaporizes the sulfur and triggers vigorous eruptions. Io's largest volcano, named Pele for the Hawaiian fire-goddess, spouts material 175 miles into space—about 10 tons of debris a second. Scientists estimate that the combined effect of all the volcanoes on Io is enough to have covered the moon's entire surface with 1,000 feet of sulfuric ash.

Europa, the next large satellite out from Io, is subjected to some of the same gravitational stresses—with different consequences. The Voyager photos of Europa revealed a surface crisscrossed with spreading networks of lines; to some of the Voyager scientists, the surface looked like a cracked egg. Europa's density is about 10 per cent less than that of Earth's moon. This implies that Europa has a rocky interior but that its highly reflective surface is composed almost entirely of water frost. Therefore, scientists familiar with satellite photographs of Earth's arctic sea ice suggested that Europa's myriad lines were cracks in the icy surface. They also speculated that an ocean of liquid water may underlie the satellite's icy crust.

To explain how liquid water could exist in the freezing void of space half a billion miles from the sun's warming rays, Steven W. Squyres and Ray T. Reynolds of NASA's Ames Research Center proposed an intriguing theory: The tidal forces that continuously reshape Io operate on Europa as well, but less strongly because of this moon's greater distance from Jupiter. It follows that the slight stretching Europa undergoes in the course of each orbit would heat its interior just enough to keep most of Europa's water in a liquid state. In this view the moon has a vast ocean capped by ice several miles thick. The Ames scientists have also wondered whether such an ocean could harbor primitive life-forms. In spite of Europa's forbidding cold, they think that this moon may be one of the most promising places in the solar system to look for signs of life.

The Voyager images presented a more conventional picture of the Galilean satellites Ganymede and Callisto. These ice-encrusted worlds are heavily cratered and less dense than the inner satellites. Callisto is the most pockmarked body in the Jovian system, its ravaged face resembling the surfaces of Mercury and Earth's moon. The abundance of craters indicates that Callisto has been geologically inactive since its formation, for Earth and other planets with active surfaces sooner or later obliterate impact craters. However, the Voyager photographs did reveal at least one kind of surface activity taking place on Callisto—a process akin to glacial flow. A number of large impact craters exhibit rims that appear worn down, and their floors are much shallower than would be expected. This lack of vertical relief suggests that Callisto's crust is soft, allowing parts of the ancient basins to collapse and blend in with the surface.

Ganymede's surface proved to be topographically more diverse than Callisto's. This largest of the Jovian satellites—its diameter is slightly greater than the planet Mercury's—shares Callisto's low density, icy crust and cratered surface. However, parts of its surface, like Europa's, are also scarred with canal-like streaks. This "grooved terrain," as the Voyager team scien-

Io's Bizarre Volcanic Eruptions

Scientists have proposed several explanations for *Voyager 1's* photographic evidence *(far right)* of volcanic activity on Jupiter's frigid moon Io. According to the prevailing theory, volcanism on Io is not produced by decaying radioactive material, as on Earth. Rather, gravitational forces constantly reshape the moon as it travels its elongated orbit, shown at right. Io is stretched as it approaches Jupiter, and it contracts as it passes well beyond the giant planet.

In an extension of this hypothesis, a group of scientists developed a model showing how the friction generated by this gravitational tug-of-war may superheat the moon's interior and trigger two types of volcanic eruption discerned on Io. Their explanation is illustrated in the cross section below.

Io's silicate crust is punctured by volcanic vents that discharge a mixture of silicates, sulfur and sulfur compounds in two types of eruptions. In the cross section below, small but long-lasting eruptions occur at left when a thin layer of molten sulfur *(red)* makes contact at a volcanic vent with sulfur dioxide *(blue lines)* that runs like ground water through the solid layer of sulfur *(yellow)*. At right, briefer but more violent eruptions occur when molten silicates from deeper levels heat and vaporize black sulfur at temperatures of 620° to 1,700° F.

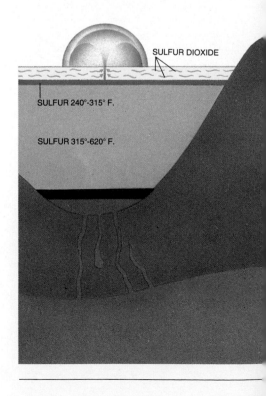

SULFUR DIOXIDE

SULFUR 240°-315° F.

SULFUR 315°-620° F.

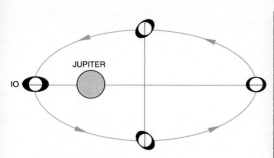

This diagram shows the gravitational and tidal forces that heat Io's interior. As Io approaches close to Jupiter, the planet's gravitational pull raises large tidal bulges on Io's surface, stretching the moon. Then Io moves away from Jupiter in an orbit elongated by the gravitational pull of two Jovian moons, Europa and Ganymede; during this phase, Io's tides shrink. The stretching and shrinking generate friction and therefore heat in Io's core. Another complex phenomenon is involved. Because Io attempts to face the geometric center of its orbit *(red line),* while its tidal bulges rise on a line toward Jupiter, the tides of Io wobble back and forth across the surface of the moon, producing even more friction and heat.

A plume of sulfur dioxide shoots skyward from Io in the *Voyager 1* photograph at right, which is falsely colored to accentuate the symmetry of the snowlike falling spray.

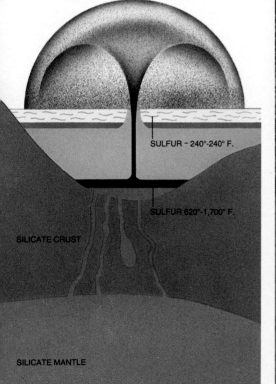

SULFUR −240°-240° F.

SULFUR 620°-1,700° F.

SILICATE CRUST

SILICATE MANTLE

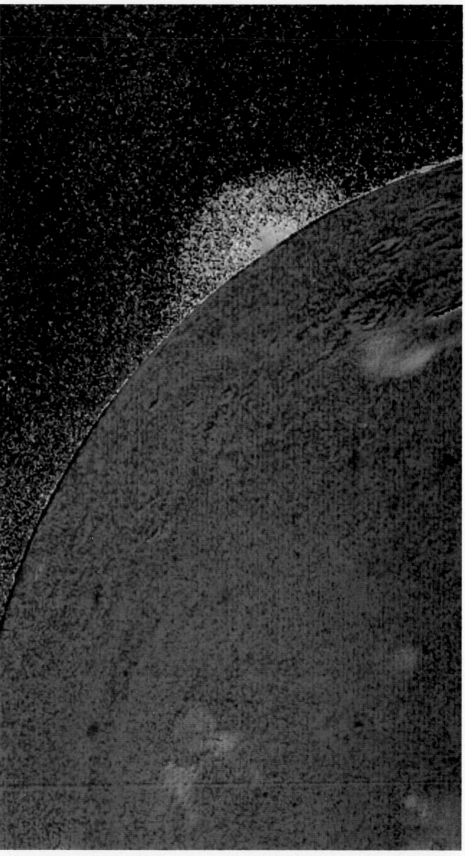

tists described it, seems to have resulted from tectonic activity during an early phase in the satellite's history. Many of the grooves run in parallel bands across Ganymede's tan-and-brown surface and may be faulted mountains similar in scale to the Appalachians of the Eastern United States. The moon also displays a number of offset groove systems. These, like the San Andreas Fault and other seismically active areas at the boundaries of Earth's tectonic plates, also seem to have been caused by faulting, with long sections of adjacent terrain moving apart from each other. This suggests to some scientists that Ganymede's icy landscape was cracked and sundered by great quakes about four billion years ago.

Leaving Jupiter behind by July of 1979, the two Voyagers pressed on toward Saturn, cruising at more than 600,000 miles a day. Even at that rate, it took them nearly a year to cover the 400 million miles to the great ringed giant. *Voyager 1* did not glimpse Saturn until November 1980, more than three years after it was launched.

In several ways, Saturn proved to be a somewhat smaller version of Jupiter. It has the same three-tiered cloud structure that obscures Jupiter, and its atmospheric movements are even more violent, recording speeds of up to 1,100 miles an hour. Though it must have lost most of its original energy long ago, Saturn — like Jupiter — generates more heat than it receives from the sun. In fact, its rate of heat output is even greater than Jupiter's. The reason for this extra energy may be that on Saturn, whose interior temperatures are lower than Jupiter's, liquid helium sinks toward the center of the planet, releasing gravitational energy that keeps Saturn warm. Also like Jupiter, Saturn has a liquid hydrogen interior and a strong magnetic field.

Despite these resemblances, the Voyager images of Saturn's collection of rings and moons provided many spectacular novelties. Morrison said that no scientist, poet or artist could have "ever imagined the awesome complexity of the tens of thousands of rings that we now recognize circling Saturn." Other unexpected features discovered by Voyager were the raylike spokes crossing Saturn's middle ring, and a pair of "shepherd" moons whose gravitational interaction keeps a narrow outer ring from flying apart. A single "guardian" moon on the outer edge of one ring performs a similar function.

The most perplexing find may have been a certain braided effect in a wispy outer ring, whose strands do not always run parallel but occasionally appear to twist or knot together. Like the radial spokes, this is an ephemeral feature whose origins still have not been satisfactorily explained. According to one theory, the braiding may result from gravitational tension imparted by the shepherd satellites, probably combined with electromagnetic forces acting on the ice particles composing the ring. Some scientists propose that electromagnetic attraction may cause the radial spokes as well.

Above all, Voyager underscored the protean nature of Saturn's rings. In analyzing the dynamics of the bodies making up the rings, Donald R. Davis and colleagues at the Planetary Science Institute in Tucson concluded in 1984 that ring particles, which vary in size from dust grains to boulders, build up and break apart over and over again in an endless game of gravitational give-and-take. Davis and his associates declared: "The traditional picture of discrete bodies, maintaining their identities over geological time scales, is qualitatively wrong. The dominant ring particles are ever-changing aggregates."

Photographed through green, violet and ultraviolet filters, Saturn reveals distinct bands in its dense, stormy atmosphere. At the equator, winds reach speeds of 1,100 mph, more than four times the velocity of Jupiter's winds.

Many of Saturn's moons are diminutive and irregularly shaped. Although astronomers refer to them familiarly as rocks, they are more likely iceballs with rocky cores. The largest and oddest rocks are a pair of coorbiting satellites, whose description derives from a kind of gravitational *pas de deux* they perform. Moving close together in parallel orbits of different speeds, the two moons periodically switch positions, with the slower assuming the other's faster orbital trajectory.

Beyond the coorbitals lies the realm of Saturn's principal moons. In sheer size, only the aptly named Titan is in a class with the Galilean satellites of Jupiter, although most have unusual features that make them fascinating objects of study. Six major moons are spherical bodies ranging from 250 to 1,000 miles in diameter, or from ⅟₃₂ to ⅛ the diameter of Earth. From calculations of their density based on size and planetary dynamics, astronomers infer that these moons are made up mostly of ice.

To the amazement of the Voyager scientists, they found that Mimas, one of the six and the smallest of the icy moons, is scarred by an enormous impact crater that spreads across nearly a third of its diameter. Planetary cartographers named the crater Herschel for England's famous astronomer. If the impacting body had been slightly larger, scientists believe, it would have destroyed the loosely cohesive moon. Indeed some scientists speculate that Mimas had previously been shattered by an impacting body, and that the orbiting debris gradually reassembled into the present lunar mass.

Although similar in size to neighboring Mimas, Enceladus appears utterly different. Its brightly reflective surface is only lightly cratered in comparison with most of the other Saturnian moons, and some of it is marbled with ridges. The sparsity of craters suggests that the surface is being continuously re-formed by some kind of geologic activity. It seems likely that Enceladus, like Jupiter's Io and Europa, is in the grip of strong tidal forces produced by the gravity of its parent planet. Enceladus travels an orbit slightly elongated by the gravitational pull of nearby Dione. As it swings toward and away from Saturn, the varying force of Saturn's gravity tugs at its interior, generating enough heat to melt some of the water ice, to soften the crust and to obliterate some of the craters. There is also speculation that Enceladus has volcanoes that spew forth water-ice droplets into space. Such icy particles may be the source of the materials composing the faint outermost ring of Saturn.

Beyond Enceladus lies Tethys, another heavily cratered iceball. Its distinguishing features are a giant impact crater similar to the one on Mimas and a sinuous north-south complex of cracks arcing across three quarters of its circumference. The crater has largely collapsed, suggesting that the interior of Tethys has been weakened and could not support the enormous basin and central mountainous structure. The branching canyon of cracks, perhaps 60 miles wide and more than a mile deep, seems to have been created by the gradual cooling of Tethys. The moon evidently froze like an ice cube, from the outside inward. The watery interior, expanding as it froze, simply burst its brittle shell.

Two tiny companion moons share the same orbit with Tethys. Known as Telesto and Calypso, they are irregularly shaped bodies only about 20 miles wide at their longest diameter. An exquisite gravitational balance holds these satellites in place 60 degrees in front of and behind Tethys. Such stable positions are known as Lagrangian points after the French mathema-

HOW A PLANET GETS ITS RINGS

Since the discovery of ring systems around Jupiter and Uranus, astronomers have come to believe that rings, once considered unique to Saturn, occur regularly as stars, planets and moons are born. The study of Saturn's elaborate rings, now known to number in the thousands, provides important insights into the mechanics governing the formation of all ring systems.

Because no one theory can explain all the phenomena observed in Saturn's system, scientists think at least three different processes, diagramed below, are involved in ring formation. A crucial factor of their theories is the Roche Limit, named for Édouard Albert Roche, the French mathematician who proposed it. The Roche Limit is the distance from a planet beyond which particles can accrete to form moons. Inside the limit, particles are prevented from coalescing by the planet's gravitational pull and by the particles' different speeds.

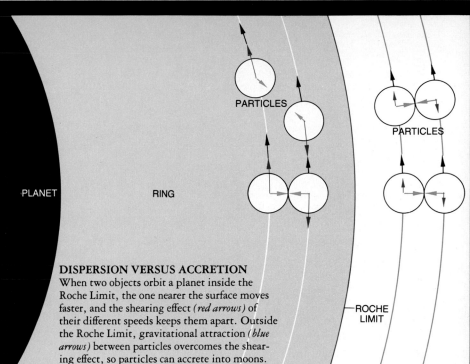

DISPERSION VERSUS ACCRETION
When two objects orbit a planet inside the Roche Limit, the one nearer the surface moves faster, and the shearing effect (*red arrows*) of their different speeds keeps them apart. Outside the Roche Limit, gravitational attraction (*blue arrows*) between particles overcomes the shearing effect, so particles can accrete into moons.

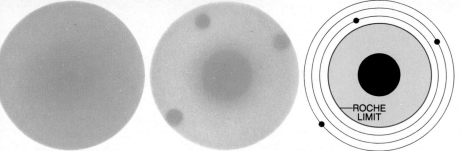

THE NEBULAR THEORY
As a vast cloud of gas and dust collapses to form a planet, a sphere of surplus debris is left orbiting the central body. Beyond the Roche Limit (*red line*) of the newly formed planet, leftover particles cluster together to form moons. Inside the limit, where particles cannot link up, constant collisions between them gradually concentrate their orbits into a thin disk revolving in the planet's equatorial plane.

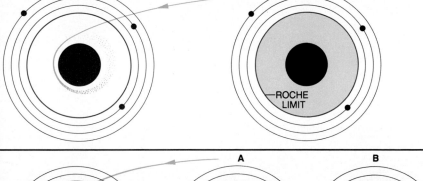

THE FOREIGN BODY THEORY
Forces within the Roche Limit can do more than prevent satellites from forming. Deep inside the limit, a planet's gravitational field, tugging at solid objects just as the moon's gravity tugs at Earth's tides, can actually tear apart a body in orbit. If a comet, asteroid or larger chunk of matter from space plunges (*green line*) into a close orbit, it can be shredded into fragments of ring material.

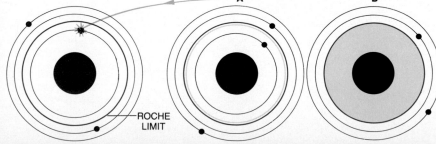

THE DEBRIS THEORY
A moon can exist within the Roche Limit if it formed while the planet was still accreting. As the planet grew, the Roche Limit expanded to encompass the moon. When a meteorite hits the moon (*green line*), chipped-off pieces can be thrown into their own orbit and form a thin ring (*blue line*) confined between two moons (*figure A*). Or the whole moon can be shattered, forming a diffuse ring (*figure B*).

SATURN

SPOKES

RING IN GAP

An Intricate Tapestry of Rings and Moons

The rings of Saturn are so thin in proportion to their width that if a scale model were to be built with the thickness of a phonograph record, the model would have to measure four miles from its inner edge to its outer rim. Since diagraming the ring system to scale is impossible, the artist's reconstruction above concentrates on illustrating key features, some of which are repeated thousands of times in the 171,000-mile width of the main rings.

The key features include spokelike accumulations of particles that cross dense areas of ring material, and a pair of small moons that hold an orbiting ringlet in shape between them. Perhaps the most conspicuous features are the several gaps in the Saturnian system.

The first gap, called the Cassini Division after the astronomer who discovered it, divides the rings into two main groups. Another gap, known as the Encke Division, was later discerned. However, recent research has revealed that both these gaps are not really empty. There are several ringlets within the Cassini, and orbiting inside the Encke is a thin, kinky ringlet that a scientist punningly named "Encke Doodle."

MYRIAD ORBITING ICE CHUNKS

Ice is the major component of the trillions of ring particles orbiting Saturn. Trace amounts of elements such as iron and sulfur must be present in the ice crystals to account for subtle variations in color, shown here greatly enhanced.

SHEPHERD
MOONS

COORBITAL
MOONS

DENSITY WAVES   MOON IN GAP   CORRUGATION WAVES   BRAIDED RIN

SHINING SPOKES OF FINE DUST

Streaks that cross many ringlets may be made of fine powder from collisions in densely packed regions. Electromagnetic forces could overpower gravity, temporarily lifting very small grains above the swarms of orbiting particles.

GUARDIANS AND SHEPHERDS

Two types of small moons embedded in the rings confine the myriad particles to a regular orbit. The particles may be held to a narrow path by the gravity of a small guardian moon in their midst, or by two shepherd satellites riding the ringlet's inner and outer boundaries.

DENSITY WAVES

Ring particles in orbit near a planet move faster than a moon farther out. As they pass the moon, the ring particles are pulled toward it and bunch up, forming so-called density waves.

HILLS AND TROUGHS IN THE DISK

Corrugation waves, rippling inward across the rings, are caused by the gravitational pull of moons that orbit Saturn outside the ring plane. These moons move up and down through the rings, ultimately pulling particles upward from above and downward from below.

GAP-CLEARING MOONLETS

In a seeming paradox of physics, a moonlet can clear a gap around its orbit by scattering particles with its gravity. Particles moving in an orbit somewhat smaller than a moonlet's travel faster than the satellite, and as they overtake it they feel its strongest gravitational effect from behind. This force from the rear tends to brake the particles, dropping them into a smaller, lower orbit. Particles orbiting just outside the moonlet are flung outward as the satellite overtakes them.

BRAIDED RINGS

At the outer rim of Saturn's main ring system is a formation that once baffled scientists with its apparent defiance of the laws of physics. The orbital paths of particles in this curious ring appear to be kinked and twisted around one another like the strands in a braid. Astronomers now believe that the ring is shaped by the gravitational influence of two moonlets shepherding its orbit.

PAIRED MOONS IN A SPACE RACE

Just beyond one of the outermost rings, a pair of coorbital satellites, one roughly spherical, the other of irregular shape, take turns catching up with each other in an intricately choreographed race that keeps them exchanging orbits without a crash. They may be the surviving fragments of a moon that was shattered long ago by an impacting body.

tician who first studied them. Neighboring Dione has a single Lagrangian moon orbiting 60 degrees in front of it.

Dione and Rhea, the next outward satellites, are distinguished by the unexplained fact that they have bright, relatively smooth hemispheres facing the direction of orbit and darker, complexly marked hemispheres on the opposite side. Otherwise, the two are typical in that they appear to have been geologically inactive through most of their history. On both moons, the most heavily bombarded areas resulted from a torrential rain of meteorites occurring early in the evolution of the solar system. Both satellites are also marked by wispy streaks that probably resulted from the same outward-to-inward freezing seen on Tethys. The wisps are especially prominent on Dione. According to the most likely scenario, Dione's interior expanded slightly as it cooled, cracking its surface. Water escaping from these cracks then froze over the face of the satellite, creating the brightly reflective swaths recorded by the Voyagers.

Titan, the next moon beyond Dione and Rhea, is three times bigger than any other Saturnian moon and second in size only to Jupiter's Ganymede among all the moons of the solar system. An estimated 55 per cent of its total mass is rock, with ice forming most of the remainder. It is unique for its reddish, smoglike atmosphere; no other moon can boast a thick atmosphere. Although Titan is only 1/50 of Earth's total mass, its atmospheric mass is actually greater than Earth's. Despite the extreme cold of its surface temperatures (on the order of −290° F.) there is speculation that hydrocarbons in Titan's atmosphere might nurture some form of primitive life.

Its tomato-soup atmosphere consists of at least 85 per cent nitrogen and possibly 12 per cent argon, a gas prominent in the primitive atmosphere of Earth and other planets. The remaining 3 per cent consists of methane and various hydrocarbons formed in the upper atmosphere by the breakdown of methane in reaction to ultraviolet sunlight and energetic charged particles. Spectroscopic studies found hydrocarbons like acetylene, ethylene and ethane as well as hydrogen cyanide, from which amino acids and other, more complex prebiological compounds can be formed.

As a test of Titan's biological potential, planetary astronomer Carl Sagan and Cornell University colleague Bishun N. Khare conducted a laboratory experiment in which a simulated Titanian atmosphere was subjected to a four-month bombardment of ultraviolet light and charged particles. The scientists succeeded in synthesizing a reddish brown powder that they called tholin, a name derived from the Greek word meaning "muddy." This substance superficially resembles the color and brightness of Titan's clouds. "Tholin seems to be a major constituent of the observed haze," said Sagan, "and thus we can make some claim to have bottled the clouds of Titan." Sagan added that tholin is "an extremely complex organic material containing many of the essential building blocks of life on Earth. Indeed, if you drop Titan tholin into water you make a large number of amino acids, the fundamental constituents of proteins."

Sagan and others hypothesize that conditions on Titan might be similar to those of a primeval Earth at the dawn of life. However, many scientists remain doubtful that any life exists on Titan and point to several factors supporting their view. For one thing, Titan's surface is almost certainly too cold for water, an essential to all known life-forms, to exist in a liquid state. For another, the Voyager mission found only a very limited supply of free

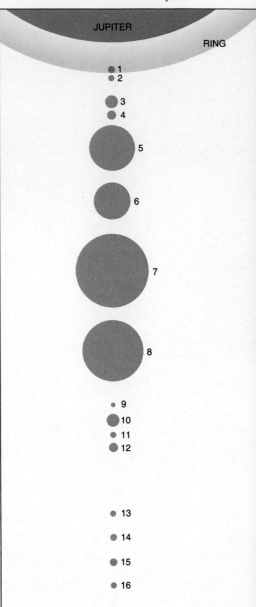

THE GIANTS AND THEIR OUTLIERS
The moons and ring systems of Jupiter and Saturn are shown below in their relative positions. Distance cannot be represented in scale; the moons would have to be too far apart.

JUPITER

RING

MOONS OF JUPITER
1. Metis: diameter, 25 miles; distance, 79,488 miles
2. Adrastea: diam., 15 miles; dist., 80,109 miles
3. Amalthea: diam., 106 miles; dist., 111,780 miles
4. Thebe: diam., 50 miles; dist., 137,862 miles
5. Io: diam., 2,254 miles; dist., 262,062 miles
6. Europa: diam., 1,950 miles; dist., 416,691 miles
7. Ganymede: diam., 3,266 miles; dist., 664,470 miles
8. Callisto: diam., 2,981 miles; dist., 1,170,585 miles
9. Leda: diam., 9 miles; dist., 6,899,310 miles
10. Himalia: diam., 115 miles; dist., 7,122,870 miles
11. Lysithea: diam., 22 miles; dist., 7,271,910 miles
12. Elara: diam., 47 miles; dist., 7,290,540 miles
13. Ananke: diam., 19 miles; dist., 12,854,700 miles
14. Carme: diam., 25 miles; dist., 13,879,350 miles
15. Pasiphae: diam., 31 miles; dist., 14,487,930 miles
16. Sinope: diam., 22 miles; dist., 14,512,770 miles

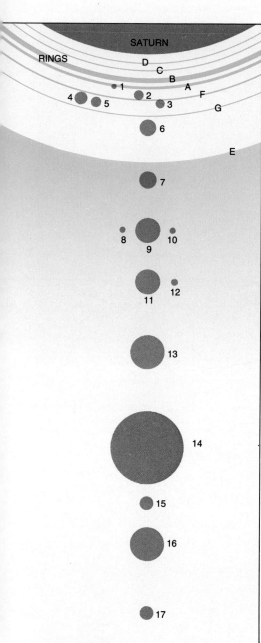

oxygen. Most of the oxygen that does exist on Titan is locked up in the ice that makes up so much of the moon.

Whether or not conditions were ever amenable to life on Titan, its atmospheric chemistry produces an abundance of carbon-based compounds that float down slowly through the smog. Two such compounds, ethane and methane, may constitute a chemical ocean covering much of Titan's surface to a depth of a half mile or more. Planetary astronomer Jonathan Lunine and others at the California Institute of Technology have proposed that the moon has a nightmarish landscape. The ocean, wrote Lunine, "would probably cover most topography that could be expected from meteorite impacts. A few islands of water ice — the 'bedrock' of Titan — might poke up. Impurities of heavy hydrocarbons would likely make the ocean a murky red color, matching the bloody and hazy red of the sky. Organic solids and tars would coat the ocean bottom and even island surfaces."

Potato-shaped Hyperion, the next moon outward from Titan, offers a striking contrast to its gigantic lunar neighbor. Scientists believe this tiny satellite was once a larger, spherical body that was broken up when struck by a meteorite. Among its puzzling features is an abnormal rotational axis. The moon is about a third longer than it is wide, and the pull of Saturn's gravity ought to orient the satellite with its long axis pointed toward the parent planet. Instead the long axis points obliquely toward Saturn, and the moon tumbles through its orbit at a variable rate. The chaotic rotation may be explained by Hyperion's gravitational interaction with Titan.

Enigmatic Iapetus, the outermost of the large icy satellites, presents a baffling two-tone appearance not unlike that of Rhea and Dione; it is as black as coal tar on one side and brightly reflective on the other. Quite the opposite of Rhea and Dione, however, the dark side of Iapetus is the hemisphere pointed in the direction of orbit. Some scientists speculate that Iapetus' blackness may result from dust swept up by the satellite in its orbital path. Others point out that most of the dark matter is concentrated on the floors of craters, and they suggest that the material may be the product of carbon-rich materials escaping from the interior of Iapetus.

If this grimy coating is dust, it may come from Phoebe, the farthest satellite in Saturn's system and a moon whose elongated orbit is steeply angled and moves in an opposite or retrograde direction from the others. Scientists theorize that tiny meteorites striking Phoebe kick off particles that drift into the orbit of neighboring Iapetus, whose leading edge collects the black powdery stuff.

Beyond Saturn, the Voyagers sped outward through a realm of dwindling twilight and yawning space. A different order of distance prevails in these dim outer reaches of the solar system. Uranus is more than twice as far from the sun as Saturn, and Neptune is half again farther than that. The distances beyond Saturn are so great that sunlight reflected off Uranus takes two hours and 45 minutes to reach the telescopes of earthbound astronomers, while the journey of light from Neptune takes more than four hours. Even when viewed through the most powerful telescopes under the best conditions, these remote planets appear only as fuzzy, blue-green blobs. And knowledge of these remote outer planets remains correspondingly vague, at least in comparison with all that science has learned about the other planets in the last decade.

MOONS OF SATURN

1. Atlas: diameter, 19 miles; distance, 85,077 miles
2. 1980S27: diam., 62 miles; dist., 86,319 miles
3. 1980S26: diam., 56 miles; dist., 88,182 miles
4. Janus: diam., 118 miles; dist., 93,771 miles
5. Epimetheus: diam., 74 miles; dist., 93,771 miles
6. Mimas: diam., 242 miles; dist., 116,127 miles
7. Enceladus: diam., 310 miles; dist., 147,798 miles
8. Telesto: diam., 15 miles; dist., 183,195 miles
9. Tethys: diam., 658 miles; dist., 183,195 miles
10. Calypso: diam., 15 miles; dist., 183,195 miles
11. Dione: diam., 695 miles; dist., 234,738 miles
12. 1980S6: diam., 19 miles; dist., 234,738 miles
13. Rhea: diam., 950 miles; dist., 326,646 miles
14. Titan: diam., 3,200 miles; dist., 758,241 miles
15. Hyperion: diam., 158 miles; dist., 919,701 miles
16. Iapetus: diam., 907 miles; dist., 2,211,381 miles
17. Phoebe: diam., 137 miles; dist., 8,058,160 miles

Uranus and Neptune are often grouped with Jupiter and Saturn as the Jovian outer planets, whose common characteristics include thick hydrogen-helium atmospheres and multiple moons or moon-ring systems. They are also large in size; they are the third and fourth largest planets in the solar system even though their masses are only about 5 and 6 per cent of Jupiter's. Like Jupiter and Saturn they are primarily gaseous. But while those gas giants are made up largely of hydrogen and helium, Uranus and Neptune also contain considerable amounts of heavier elements, including oxygen, nitrogen, carbon, silicon and iron.

Though scientists have been unable to see through the opaque atmospheres of Uranus and Neptune, they have developed theoretical models of the planets' interiors based on the physics governing the behavior of materials under the conditions of cold and pressure prevailing there. The models indicate that each of the planets has a surface of compressed hydrogen and helium, which lies over a mantle of liquid or frozen methane, ammonia and water, which in turn surrounds a rocky or liquid core of metals and silicates. Observations suggest that Neptune has an internal heat source. Perhaps it is still cooling off, losing heat generated by gravitational contraction during the formation of the planet. Whatever caused its heat, Neptune remains at about the same temperature as Uranus—a brisk −357° F.— despite its greater distance from the sun.

Surely the oddest feature of Uranus is its axis of rotation. Its poles are tilted 98 degrees from the vertical, so that it circles the sun lying on its side. Since the planet takes 84 years to complete one orbit, its polar regions are alternately exposed to 42-year periods of sunlight and of darkness.

Uranus has nine known rings, the first of which was discovered by telescope as late as 1977. Although the rings cannot be seen directly, they were detected by occultation—their dimming effect on light from a star as Uranus crossed in front of it. The rings are extremely narrow; the largest is about 60 miles across, and most of them are only about six miles wide. They are also very dark, indicating the presence of carbon-rich materials such as silicates. At least six of them are out-of-round. Their curious shape and crisp, well-defined edges imply the presence of shepherd moons that, like Saturn's, maintain the rings in their eccentric orbits.

Five satellites, discovered between 1787 and 1948, are known to orbit Uranus. Their names, drawn principally from characters in Shakespeare's *A Midsummer Night's Dream* and *The Tempest,* are Miranda, Ariel, Umbrel, Titania and Oberon. Spectrographic studies of these far-off moons show that they are iceballs similar in size to most of Saturn's satellites, but with more dust and rocky debris mixed in.

The two known moons of Neptune, Triton and Nereid, are remarkably eccentric. Instead of conforming to the norm for the big satellite systems by orbiting within the equatorial plane of the parent body and in the same direction of rotation, Neptune's anomalous moons are highly inclined from the equator. Also, Triton's orbit is retrograde (in opposite direction from the planet's rotation), while tiny Nereid's orbit is a very elongated ellipse. In addition, mathematical analysis of Triton's orbit indicates that the moon is slowing down and slipping dangerously close to Neptune's gravitational grasp. Scientists predict that relatively soon—perhaps even in the next 10 to 100 million years—the moon will be torn apart by increasingly strong tidal forces. When this happens, the fragments may form a new ring.

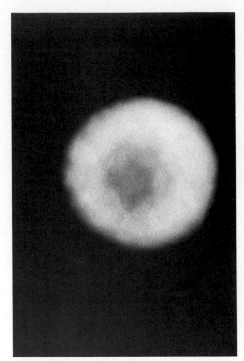

Uranus, its dark bulk encircled with bright clouds of methane ice crystals, is captured in a picture taken through a telescope at the University of Arizona at Tucson. The telescope was equipped with a special device that captured more light and thus more details than the most sensitive of standard optics.

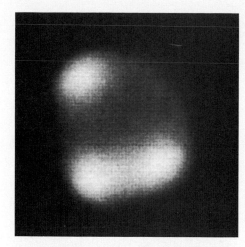

Neptune's blurred disk is partly hidden by three bright patches of methane clouds, each about the size of Earth. The picture, taken at the Carnegie Institution's Las Campanas Observatory in Chile by the same method as the one at left, is more distorted because Neptune is about a billion miles farther away than Uranus.

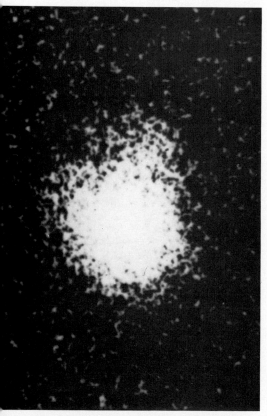

This historic picture of Pluto, heavily distorted by tremendous distance (the planet lies about one billion miles beyond Neptune), reveals that Pluto has a moon, visible as a lump at the upper right of the planet's fuzzy outline. The image, recorded by an electronic camera, is actually less distorted than any taken by standard optical cameras.

Triton is intriguing for at least one more reason. In 1983, University of Hawaii astronomers observing it with infrared sensors made a startling discovery: Triton has an atmosphere, probably of methane, and it may be covered by an ocean of liquid nitrogen with solid methane icebergs in it.

Pluto, nearly 3.7 billion miles from the sun and 900 million miles beyond Neptune, barely deserves planetary status. Repeatedly, new discoveries have forced astronomers to lower their estimates of its size and mass. Some suggest that it was once a moon of Neptune. In theory, a close encounter with the Neptunian moon Triton could have kicked Pluto into its own orbit around the sun and thrown Triton into its retrograde orbit. Other scientists now believe Pluto should be demoted to asteroidal status, for it appears to be an icy rock with a diameter of only 1,864 miles and a mass about $\frac{1}{400}$ that of Earth. Yet despite its small size, it seems to have a tenuous atmosphere of methane and some other heavy gas, as well as a surface of frozen methane, whose temperature is no greater than $-350°$ F.

Whether Pluto is a lost Neptunian moon or a displaced asteroid, its tiny size and eccentric orbit strongly suggest an origin different from that of the other planets. Its orbit departs at the considerable angle of 17 degrees from the general orbital plane of the planets. The orbit is also highly elliptical, which sometimes brings Pluto inside the orbit of Neptune. Pluto reached that segment of its orbit in 1979 and will remain in it until 1999; thus, for the time being, it does not even have its usual distinction of being the outermost planet — if it is a planet. The puzzling little body takes an astonishing 248 years to complete one orbit.

Pluto has as company in its distant orbit a single satellite discovered in 1978 by James W. Christy of the U.S. Naval Observatory. This most distant moon is also the largest moon relative to the size of its parent body; it has a diameter that is almost 40 per cent of Pluto's. The moon also revolves very close to Pluto — only about 10,500 miles away — and races through its little orbit in just 6 days 9 hours 22 minutes. Christy named the satellite Charon, after the boatman who ferried the dead across the River Styx to Hades, the underworld domain of Pluto.

In a curious way, the discovery of Pluto's moon has revived speculation that the solar system has an undiscovered 10th planet — Planet X. That possibility had been all but dismissed by the exhaustive astronomical search that Tombaugh conducted after he discovered Pluto more than 50 years ago. But recent studies of the movement of Pluto's moon showed that Pluto has too little mass and size to justify an earlier assumption that it caused minor discrepancies in the orbits of the planets from Jupiter outward.

These orbital discrepancies could be accounted for by the existence of an unseen 10th planet possessing certain essential attributes. Such was the conclusion drawn by scientists at the U.S. Naval Observatory, whose leading investigator, Thomas Van Flandern, explained the requisite characteristics of Planet X at a meeting of the American Astronomical Society in 1981. He said that the putative planet would have to have a mass of two to five times that of Earth, and that it would have to orbit between 50 and 100 astronomical units (4,650,000,000 to 9,300,000,000 miles) from the sun, in a highly inclined plane similar to that of Pluto's orbit.

Clyde Tombaugh attended that meeting at the age of 74, and he conceded that a planet as dim and distant as the one Van Flandern postulated might have escaped his notice. The Naval Observatory's search goes on. Ω

133

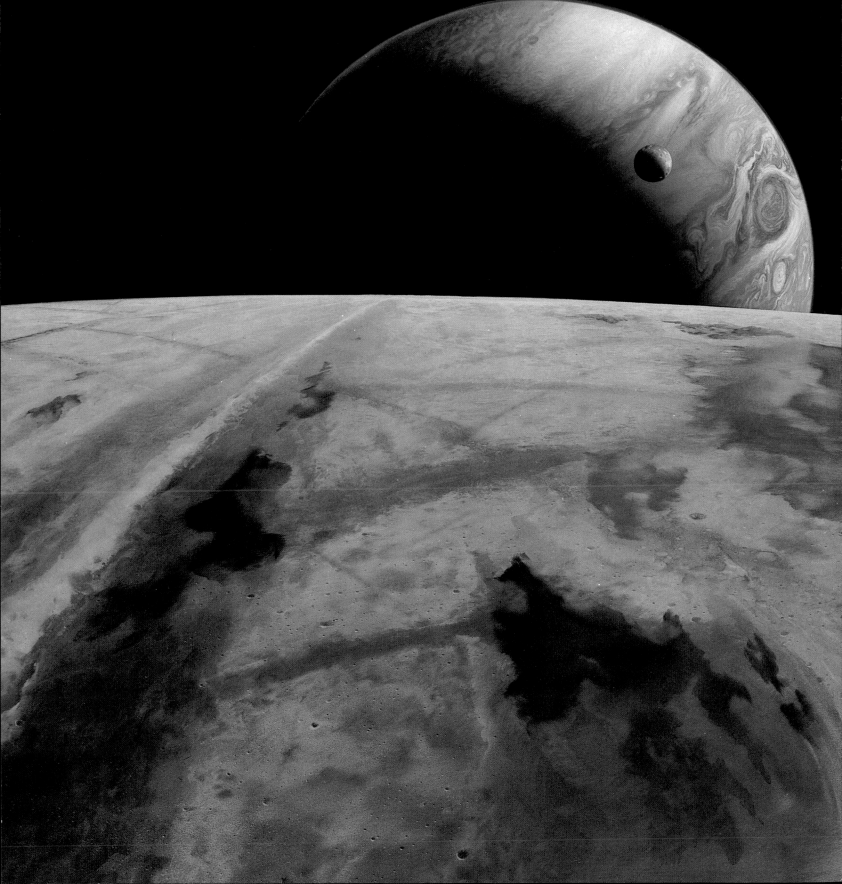

As the two Voyager spacecraft swept past the moons of Jupiter and Saturn, they photographed a startling variety of terrain: surfaces scarred by impacting bodies, glazed by flowing ice and drenched in volcanic sulfur. Some of the Voyagers' photographs appear on these pages with close-up paintings based on the latest scientific findings.

Jupiter's four planet-sized moons run a gamut of geological phenomena. Callisto is pocked with craters, indicating that it has been dead for at least four billion years. But Ganymede's cratered surface has been resculpted by internal forces that buckled its crust. On Europa, ice that may have been squeezed up from the interior has filled cracks and partially covered craters. And Io, where bizarre volcanoes spout clouds of sulfur compounds, is by far the most geologically active body in the solar system.

In this painting of Europa's glazed surface, a low frozen ridge runs to the horizon, where hover gigantic Jupiter and the moons Io *(left)* and Ganymede. On Europa, water ice may have oozed like lava into cracks and craters, making this the solar system's smoothest moon. Nevertheless, Europa looks like a badly cracked eggshell in a Voyager photograph *(below)* taken from 150,600 miles.

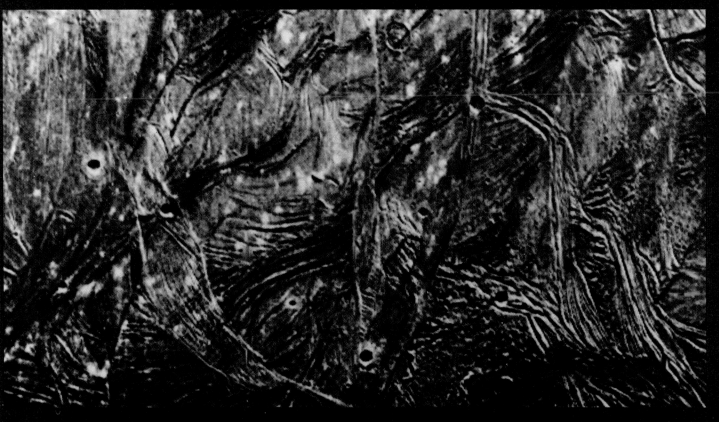

A large dark area on Ganymede bears scars of space debris that crashed ages ago. The white swatches are sprays of ice ejected by later impacting bodies.

Parallel ice ridges and grooves crisscross the surface of Ganymede. These features were probably caused by faulting in the moon's crust.

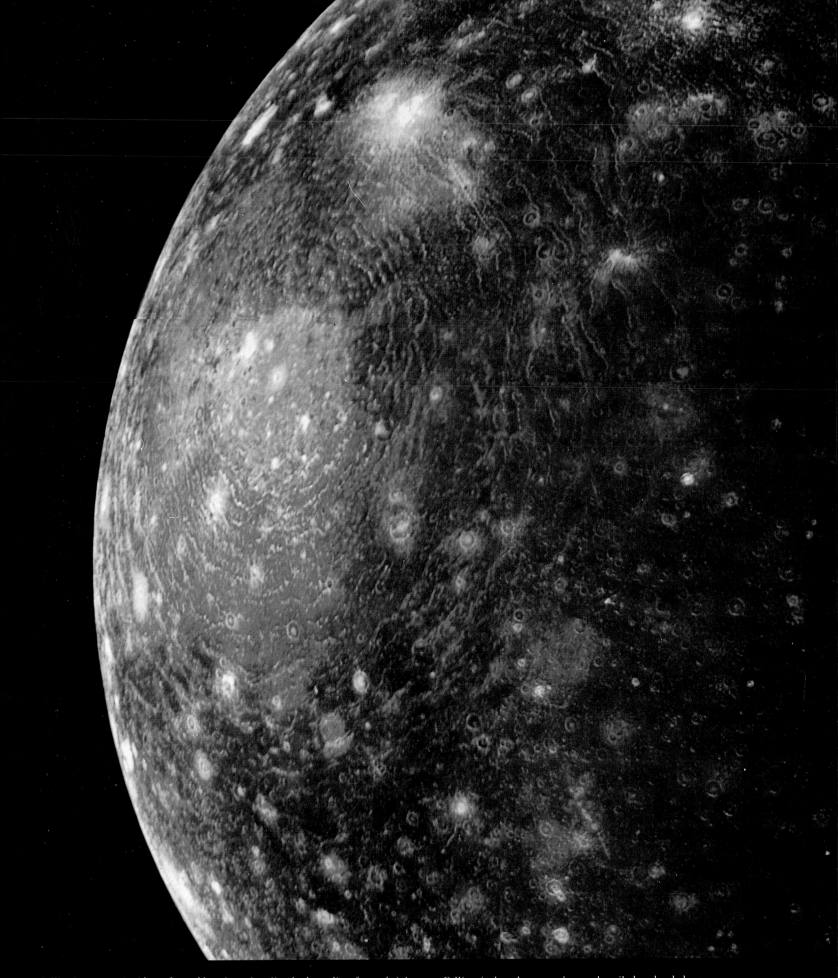

Callisto's concentric ridges, formed by a huge hurtling body, radiate from a bright core. Callisto is the solar system's most heavily bombarded moon.

The painting at right shows wispy plumes of
sulfur-rich vapors shooting hundreds of miles
upward from volcanic vents on Io, then falling back to the varicolored surface. And, at left
center, an island of solid sulfur floats on
a molten lake. The sun is flanked by a pair of
mock suns, or sundogs, formed as light is
refracted, or bent, by sulfur dioxide ice crystals
in Io's volcanic plumes. In the photograph
below, the lava lake is the dark feature at lower
right; directly above it, the double volcanic
plumes appear as bright fans connected by an
elongated deposit of sulfur or silicate.

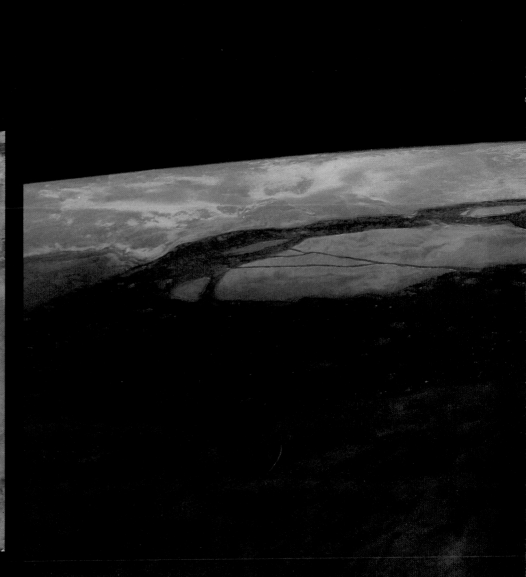

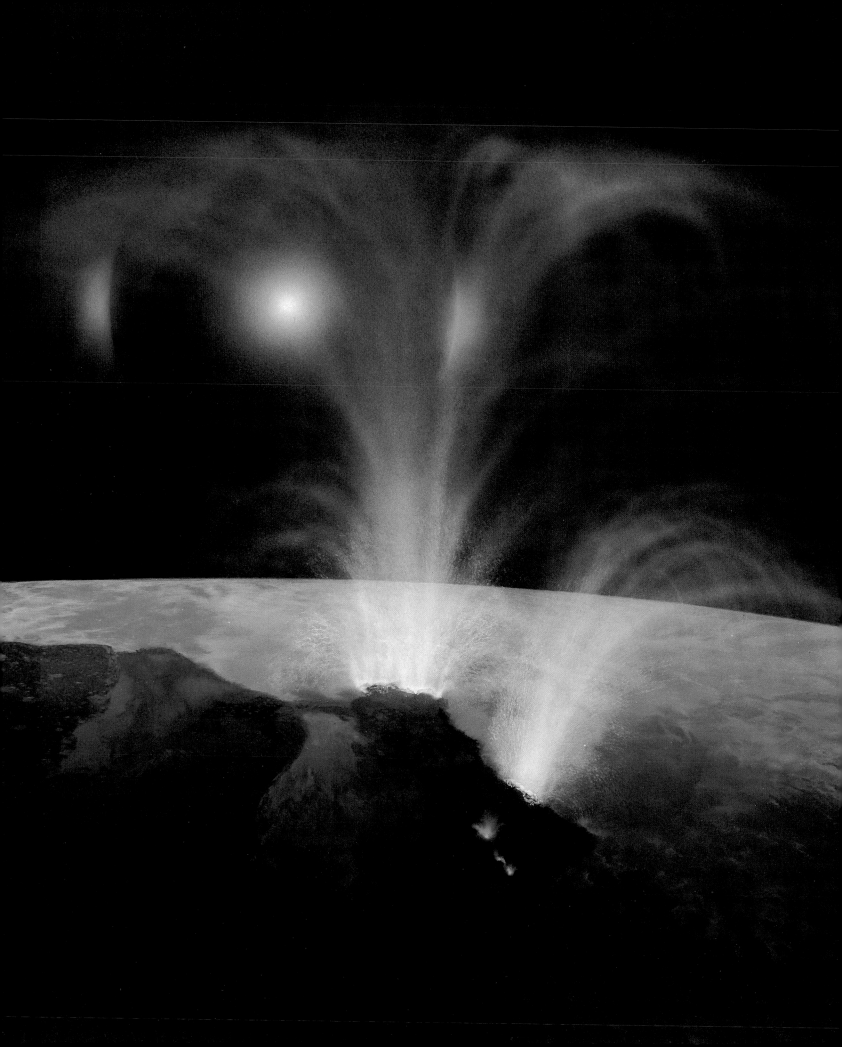

Saturn's Icy Moons

Scientists expected the icy moons of Saturn to be too small and frigid to support geological activity. And so it seemed to be at first glance: Saturn's heavily cratered moons were apparently locked in a permanent deep freeze. On Mimas, for example, an immense crater and its monolithic central peak have obviously remained unaltered since the original impact more than four billion years ago, testifying to the steely rigidity of ice in the Saturnian system.

However, a closer examination revealed several sparsely cratered areas on Tethys, Rhea and Dione, implying that some internal force had repaved their surfaces at a later stage. The same phenomenon is responsible for even more dramatic results on Enceladus, which has an enormous expanse of completely craterless terrain. Next to Io, in fact, Enceladus may be the most geologically active moon in the solar system.

Silhouetted against Saturn's rings, an icy peak 3,600 feet higher than Mount Everest rises from Mimas' gargantuan impact crater in the painting at right. Six-mile-high cliffs, perhaps the highest in the solar system, rim the crater. The crater is one third the diameter of the moon, as seen in the photograph below. Some scientists believe that if the impacting body had been slightly larger, it might have shattered Mimas.

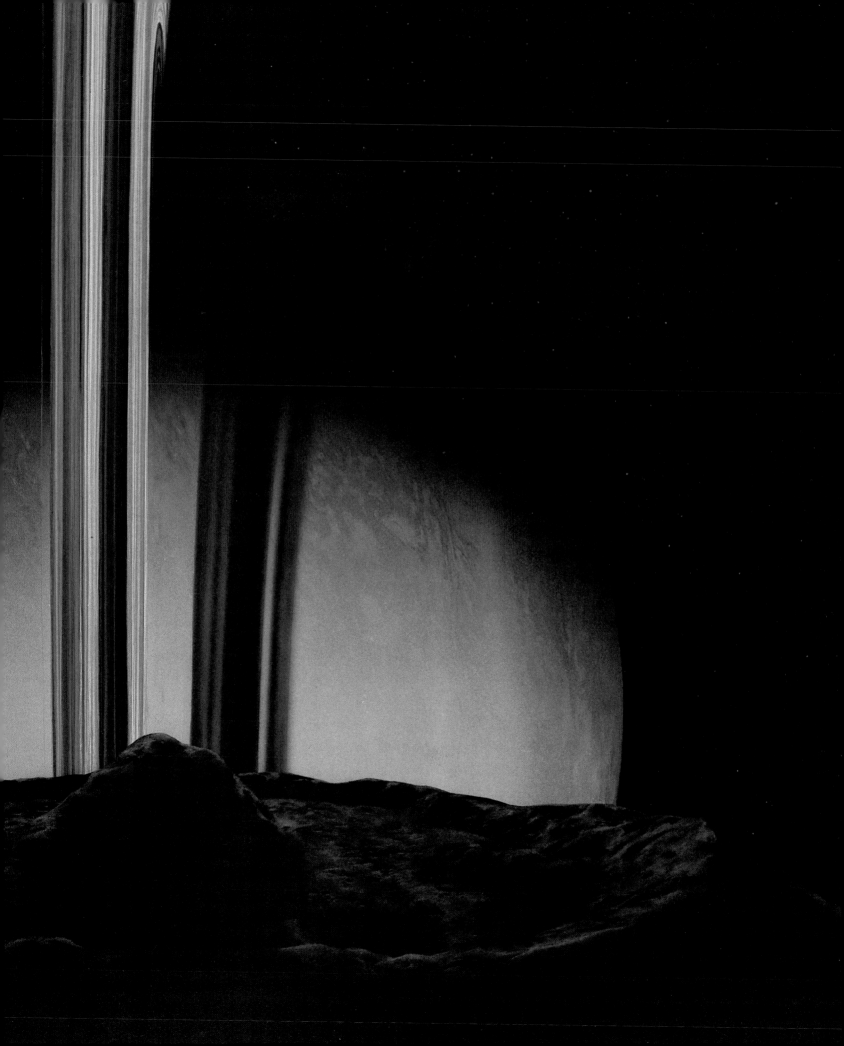

On bleak Dione, a sinuous fracture line *(top)* and bright, wispy streaks of ice *(far right)* extend far across the moon's crater-pocked terrain.

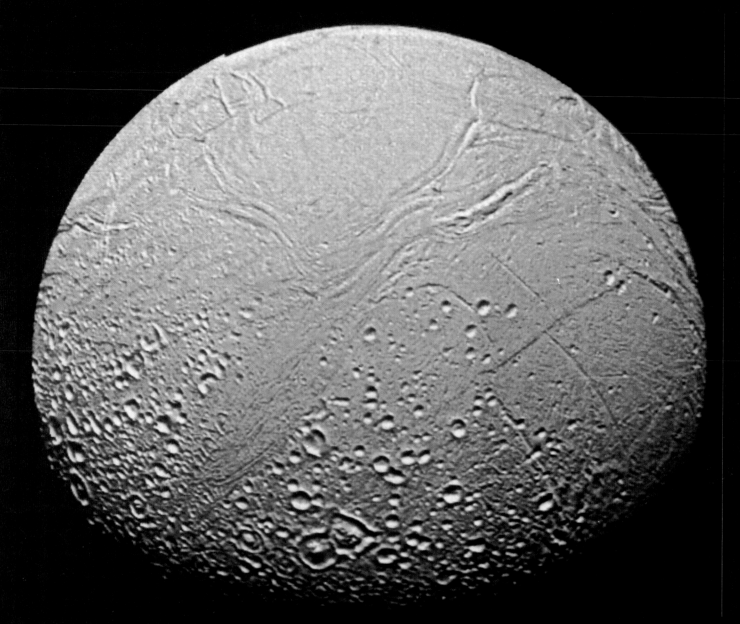

Enceladus' heavily textured surface displays smooth and ridged plains *(top)*, lightly cratered terrain *(middle)* and heavily cratered areas *(bottom)*.

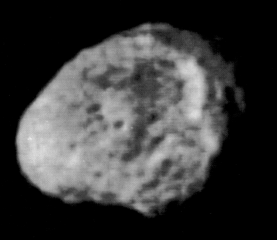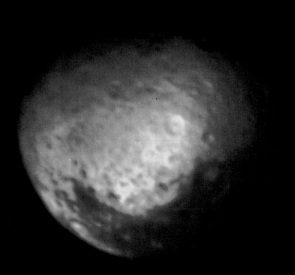

Little Hyperion *(left)* is thought to be a chunk of a shattered moon. Iapetus is a divided world — half bright ice and half dark material.

In the painting at right, the faint, distant
sun struggles to break through Titan's smoggy
atmosphere, while a constant drizzle of rich
organic compounds falls on swampy land masses
and a globe-girdling ocean *(background)*.
The ocean consists of liquid ethane and methane,
thought to be end products of photosynthesis; this, together with the rainfall of organic
compounds, might suggest the possibility of
life-forms — were it not for Titan's intense cold
and lack of free oxygen. Liquid ethane also
forms pools in the mucky, cratered terrain. In
the photograph below, a blue band of haze,
its color exaggerated to emphasize Titan's several
atmospheric layers, floats above opaque orange clouds that completely obscure the surface.

RELICS FROM TIME ZERO

For a lucky few of the giant reptiles, the end came with devastating swiftness. They perished in an instant, crushed by the shock wave from a six-mile-wide body—a comet or perhaps an asteroid—that roared out of the sky 65 million years ago. But for the vast majority of dinosaurs, death came slowly, during a long catastrophe that followed the thunderous crash.

The comet—perhaps 100 times the size of the one that exploded over Tunguska in Siberia in 1908—slammed into Earth with the force of 100 million million tons of TNT. The impact hurled more than four trillion tons of dust into the atmosphere. Carried on the jet stream, the dust spread worldwide, blocking sunlight for more than six months; several years passed before all the dust settled out of the atmosphere. The long, cold darkness suppressed plant growth and disrupted the food chain, leading to wholesale extinctions. Not only the dinosaurs died off; with them, more than 90 per cent of Earth's animal and plant species ceased to exist.

This grim scenario was proposed in the late 1960s and was quickly dismissed as wild speculation. But recently it has been transformed into an arguable theory by supporting geological finds of extraterrestrial materials. Spirited debates have taken place, and proponents of the theory have extended it with even more controversial hypotheses about other great dieoffs. Thus, regardless of its merits, the extinction theory has had a beneficial effect: It has injected new vitality into the neglected study of comets, asteroids and meteorites—the small bodies of the solar system.

A happy concatenation of events has swelled the renewed interest. There is the much-heralded return of Halley's Comet, whose 76-year-long orbit would bring it back into the inner solar system in late 1985. In anticipation of the visit, Japan, the Soviet Union and a consortium of European nations laid plans to send an unmanned spacecraft to rendezvous with the comet and observe its progress. And plans have been made to expand scientific knowledge of asteroids. NASA proposed, for a mission later in this century, to send a manned spacecraft to investigate the Apollo asteroids (named after the Greek sun-god), a swarm of small rocky bodies that regularly cross the orbits of Mars and Earth in the course of their own orbit around the sun.

Scientists new to the study of small bodies are finding that it is—as it always has been—a difficult field. Comets—those fleeting chunks of ice and rock from far-off space—are so small that they cannot be seen until they enter the inner solar system, and the source of many has been determined only by calculating the outer limit of their orbits from the loop they make around the sun. Asteroids are typically larger than comets; they range

Stars appear to swing in circles above an observatory dome during this 10½-hour exposure, taken on Siding Spring Mountain in Australia. Actually, it is Earth's rotation that traces the pattern in the night sky, while people carrying flashlights draw the squiggles across the bottom of the picture.

upward in size to several hundred miles in diameter. Even so, they are hardly easier to observe. The great majority of them occupy the asteroid belt between Mars and Jupiter, a belt that starts 200 million miles from the sun and extends 100 million miles farther. At such distances, even the largest asteroids are mere pinpoints of light.

Fragments of both asteroids and comets occasionally reach Earth. Such objects are technically defined as meteoroids while they speed through space, as meteors when they make their flaming entry into Earth's atmosphere, and as meteorites once they hit the ground. But even though meteorites have been subjected to careful analysis, their chemical composition often fails to prove conclusively whether they came from an asteroid or a comet. In fact, the efforts of scientists to distinguish between small bodies by their composition and origin have produced quite the opposite effect: The mounting weight of evidence tends to minimize the differences between comets and asteroids. "We are finding," summarized astronomer William K. Hartmann of the Planetary Science Institute, "more of a continuum among the smaller bodies, rather than distinct categories."

In a curious way, this ambiguous conclusion strengthens a basic concept about small bodies. Comets and asteroids alike are relics of Time Zero — a term space scientists use to signify the very beginning of an event, a process or a phenomenon. They are leftovers from the whirling nebula of gas and dust that spawned the solar system 4.6 billion years ago, and their chemistry provides clues to the process by which planets and moons were formed.

Comets exist in two forms. In their familiar form, visible to the naked eye, their flamelike tails — the product of solar heat vaporizing their various kinds of ice (including water, methane and ammonia ices) — stretch for millions of miles and have made them objects of wonder and dread. But as scientists have learned after arduous labors, comets also exist in a second state, visible only by telescope and unrecognized for centuries. Once a comet has all its volatile ices burned off, it is reduced to a rocky nucleus virtually indistinguishable from an asteroid. This confusing resemblance is only one of several critical problems that have haunted the study of small bodies.

For all practical purposes, serious research into the origin of comets began in 1577, when the Danish astronomer Tycho Brahe contradicted the contemporary view that comets originated within Earth's atmosphere and declared that they traveled in paths far beyond the moon. Toward the turn of the 18th Century, Isaac Newton deduced that comets have elongated elliptical orbits, and Edmond Halley, for whom the best-known comet is named, tracked the orbits of 24 comets.

After that, however, the next great landmark in comet research did not come until 1950, when Dutch astronomer Jan H. Oort postulated the existence of a cloud of comets, later known as the Oort cloud, at the extreme outer reaches of the solar system. There, the vast majority of comets — and scientists estimate that the solar system holds at least 100 billion — move about in a spherical area of space thousands of times farther away than Pluto. Held in the sun's weak gravitational embrace, comets in the Oort cloud circle in immense lazy orbits that take thousands of years to complete.

Occasionally, a gravitational disturbance — caused by a passing star, for instance — nudges a few comets out of their Oort-cloud orbits and sends them plunging toward the sun. Some take up new, elongated elliptical

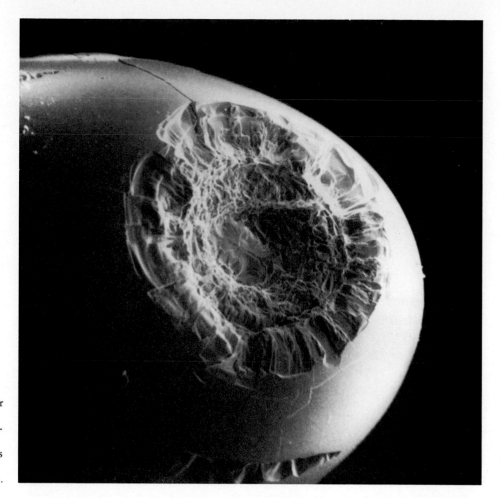

Magnified 100 times, a tiny grain of glassy material from Earth's moon reveals a microcrater gouged out by the collision of a still smaller particle of cosmic dust. The concentric and radiating fractures are similar in configuration to those found in impact craters many miles across, typifying the physical patterns that recur on every scale throughout the solar system.

orbits that loop the sun and return. More than 100 comets complete these elongated orbits in 200 years or less; they are called short-period comets. Scientists have also tracked 500 comets with longer periods.

As comets cut through the solar system and their volatile ices vaporize under the sun's warming rays, they leave behind a trail of dust particles. Whenever Earth in its orbit intersects such a trail, some of the particles fall into the atmosphere and incandesce in a spectacular display. These meteor showers are usually named after the constellation from which the meteors seem to appear. Well-known examples include the Lyrids shower (from Lyra) in April and the Leonids (Leo) in November.

Because comets are loose and fragile bodies, with only small pieces of rock embedded in them, most of the debris responsible for meteor showers is consumed by friction-generated combustion as the debris passes through Earth's upper atmosphere. The small quantity that survives the searing heat of entry usually joins the miscellaneous microscopic debris drifting in the stratosphere miles above Earth's surface. These stratospheric particles are sometimes called Brownlee particles after their discoverer, American astronomer Donald Brownlee. In general, cometary material is seldom large enough to survive the descent, leading to the logical assumption that most meteorites of any size come from asteroids. Some meteorites, weighing as much as 60 tons and measuring six feet across, have carved out big impact craters, one of the largest being the 4,000-foot-wide Meteor Crater in Arizona. But the bodies that formed them were invariably fragmented and scattered, and thus have not been analyzed in their pristine form.

Perhaps the best place for meteorite collecting is the frozen wastes of Antarctica, where glaciers have preserved the rocky remnants for hundreds of thousands of years. The 1969 discovery of fields of meteorites sprinkled

through the icecap like placer gold in mountain streams has given scientists more than 6,000 specimens to study. Elsewhere, about a dozen sizable meteorites are collected annually.

Careful chemical analysis of many specimens was to produce equivocal results. But it did establish clearly that meteorites fall into three main types: so-called "irons" (5.7 per cent of those collected), which contain a variety of metallic materials; very rare "stony irons" (1.5 per cent), with a rough balance between rocky and metallic constituents; and, by far the most common (92.8 per cent), "stones," which are rich in silicate and carbon compounds. All three types are extremely old, especially a carbon-rich subclass of the stony group known as chondrites. Some examples of chondrites have been radiometrically dated back four billion years and more. Scientists speculate that chondrites may be chemically identical to many of the planetesimals, the ancient building blocks of the planets.

Perhaps the oldest and most informative meteorite find was a chondrite that fell to Earth on the night of February 8, 1969, near the village of Pueblito de Allende in northwestern Mexico. As recalled by frightened witnesses, the Allende meteor burned across the sky in a brilliant pulsating blue-white flare. A chain of detonations cracked the nighttime silence — sonic booms. Then the blazing object broke into pieces, scattering several tons of debris in a spectacular cascade of fireworks. When scientists collected and analyzed the Allende fragments, they found evidence of aluminum-26, a rare and short-lived aluminum isotope that is known to be created in supernova explosions, the violent death spasms of certain huge stars. This link to a supernova, together with radiometric dating that placed the age of the Allende fragments at 4.6 billion years, added support to the prevailing theory, disputed by scientists who advocate several other theo-

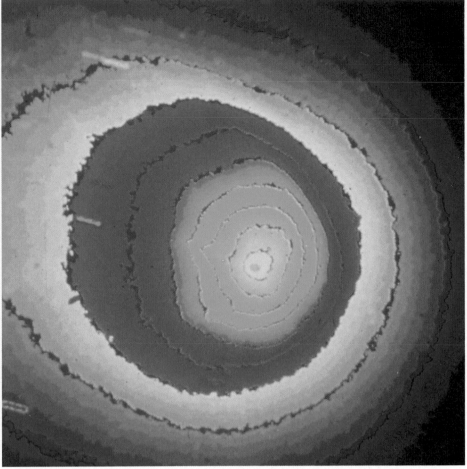

Halley's Comet, passing Earth on its 1910 visit *(left)*, streaks across the night sky against a backdrop of stars blurred by the tracking motion of the camera that took the picture. The photograph, one of four taken with black-and-white film from an observatory in Egypt, has warm tones because it was recently re-photographed with color film. In the image at right, the four 1910 photographs have been superimposed and computer-enhanced to show, through an arbitrary color code, that the comet's head, or coma, is brightest at its center, where gas and dust from the nucleus evaporate under the sun's heat.

ries, that a supernova explosion started the birthing of the solar system.

While meteorites that came from asteroids have undoubtedly been found since Classical times, the first asteroid to be identified as such was not discovered until 1801. On New Year's Day of that year the Italian astronomer Giuseppe Piazzi noticed what seemed to be a faint star in the constellation Taurus. When the object's position changed over the next several nights, Piazzi decided it was something within the solar system, perhaps a comet. Unfortunately, Piazzi fell ill and failed to make enough observations to provide an accurate calculation of orbit, and by the time word of the new object reached other astronomers it was too close to the sun to be seen.

Scientists were excited about Piazzi's discovery, which most of them assumed was a new planet. But this enigmatic object remained lost until the challenge of locating it was taken up later in 1801 by a young German genius named Johann Friedrich Gauss, who has since been ranked with Archimedes and Newton as one of the greatest mathematicians of all time. Gauss applied his own mathematical methods to Piazzi's data and predicted where the lost body would be located on the last day of the year. And there an astronomer found it, in what is now known as the asteroid belt, about 250 million miles from the sun. Piazzi named the object after Ceres, the tutelary goddess of Sicily, where he had been when he first saw it 364 days before. Ceres turned out to be only 600 miles in diameter, making it the smallest heavenly body discovered up to then. (It is also the largest of the asteroids, all of which would add up to considerably less than the mass of Earth's moon. Later evidence suggested that the asteroids never accreted into a single planet because of the gravitational tug of Jupiter.)

In the six years after the discovery of Ceres, astronomers found three similar bodies in the asteroid belt — Pallas, Juno and Vesta. By the time

Vesta was found, Gauss had become so skilled in his calculations that he was able to work out its complete orbit within 10 hours of receiving the observational data from the astronomer who discovered it.

Systematic searching with more powerful instruments turned up several hundred asteroids by the start of the 20th Century. Twenty-three years later, astronomers catalogued the 1,000th asteroid, which they named Piazzi after the discoverer of the first. Since then, the number catalogued has grown to nearly 3,000, and a conservative estimate places at 20,000 the grand total of asteroids in the solar system. Among these are the small group of Earth-approaching Apollo asteroids, discovered in 1932, and the Trojan asteroids that escort Jupiter, the first of which was found in 1906.

Some of the most important recent data on asteroids has come from the Infrared Astronomical Satellite, or IRAS for short. IRAS, a joint British, Dutch and American project, was built by NASA and launched in January 1983 into an orbit 560 miles above Earth. Free of atmospheric distortion, IRAS scanned space looking for cool to warm bodies and particles radiating low-energy infrared light. Its eight-month mission was extended to 10 and produced what a mission scientist called "some of the most revolutionary observations in the history of space-based astronomy," including tantalizing data on the evolution of stars and, possibly, new solar systems (*page 164*).

IRAS discovered two new asteroids and seven new comets. One of the asteroids, designated 1983TB by the standard discovery code, made a substantial contribution to a theory that was gaining considerable strength: Many bodies that looked and behaved like asteroids were in fact the rocky nuclei of burned-out comets. To all appearances, 1983TB was a normal Apollo asteroid. But its orbit resembled the elongated orbits of comets and brought it to within 12 million miles of the sun — significantly closer than

A long, blazing streak marks the path of a meteor across the sky over Finland. The photograph was taken during the Upsilon Pegasid meteor shower, which peaks early each August. In three nights in 1978, meteors appeared at the rate of about 20 per hour.

This six-pound chondritic meteorite, consisting chiefly of stony material, is the most common type found on Earth. It was recovered from under a dining-room table after it had plunged through the roof of a house in Wethersfield, Connecticut, in November 1982.

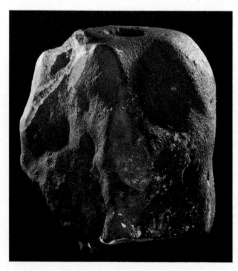

152

the other Apollos. Scientists had begun to think that some of the Apollos were actually dead comets, and 1983TB seemed very likely to be one.

Five months after 1983TB's discovery, astronomer Fred L. Whipple of the Harvard-Smithsonian Center for Astrophysics clinched the cometary identification by linking that body with a meteor shower. Although nearly all meteor showers have been connected with known comet paths, one prominent shower — the Geminid of mid-December — had previously eluded any cometary link. But Whipple's calculations showed that the Geminids occur whenever Earth crosses the orbital track of 1983TB, leading to the conclusion that this supposed asteroid is in fact a dead comet.

Early in 1984, Eleanor Helin and R. Scott Dunbar of the California Institute of Technology discovered another asteroid that turned out to be a dead comet. Asteroid 1984BC had the elliptical orbit of a typical short-period comet. Its small size — less than three miles in diameter — also fit the typical profile of a comet nucleus. And its darkly carbonaceous surface marked it as a cometary object born in the solar system's icy outer realm.

Meanwhile, spectroscopic studies of the asteroid belt showed that the chemical composition of individual asteroids reflects the same gradation of materials that distinguished the inner and outer planets during their formative stage 4.6 billion years ago. There are three main types of asteroids. About 70 per cent of those occupying the inner part of the belt are stony, or S-type, objects; like the nascent inner planets, whose gases had burned off because of their proximity to the sun, they consist chiefly of silicate minerals. Also evident here are M-type asteroids, which have a high metallic content, mostly nickel and iron. In the middle and outer parts of the belt are C-type asteroids, composed chiefly of carbon materials. These carbonaceous asteroids, like the gaseous and icy outer planets, condensed from the solar system's spawning cloud at lower temperatures than the inner planets.

Reviewing their chemical analysis of meteorites, scientists saw that the gradation between iron and stony examples corresponds to the silicate-to-carbonaceous sequence in the spectroscopic studies of asteroids. The burden of evidence suggested that almost all meteorites are fragments from collisions among asteroids in their home belt between Mars and Jupiter.

Apparently the gravity of Jupiter disturbs the orbits of the main-belt asteroids and causes collisions at speeds of 10,000 mph or more. These crashes break down or chip away at the rocky bodies and provide the solar system with a steady supply of asteroidal debris. Scientists now believe that some of these fragments drift into regions of gravitational instability in the asteroid belt known as Kirkwood gaps (after their discoverer, astronomer Daniel Kirkwood). From there they are easily nudged out of the belt and may take up an orbit that crosses Earth's route around the sun.

Such studies have gradually blurred the distinctions between asteroids and comets. In addition, recent research suggests that many small bodies have changed roles and places. Several smaller moons — among them Mars's Phobos and Deimos and Saturn's Phoebe — may have been asteroids that were kicked out of the main belt by some gravitational disturbance and then were captured by the gravity of their present host planets. Tiny Pluto may have started out as an asteroid and passed through a period as a moon of Neptune before jumping into its present planetary orbit around the sun. Some scientists think that a large number of planetesimals, left over after multitudes of them had accreted to form the various planets, later found their way into the main asteroid belt, where they joined planetesimals that were formed there. "My own opinion," says Eugene Shoemaker of the U.S. Geological Survey, "is that the asteroid belt is a zoo of rare animals captured from all different parts of the solar system."

The confusion over classifying comets and asteroids was shared by a distinguished scientist who played a key role in redeeming the hypothesis that a crashing body led to the extinction of the dinosaurs. Nobel physicist Luis W. Alvarez, along with his geologist son Walter and two of their associates at the University of California at Berkeley, declared at first that an asteroid had caused the great die-off. Later they changed their minds and said it was a comet. Despite their understandable uncertainty, their solid research lent credence to the theory and stimulated the study of small bodies.

Alvarez and his colleagues discovered, in sites in Italy and Denmark, that clay sediments deposited at the temporal boundary between the Cretaceous and Tertiary geological periods contained amounts of the trace element iridium in concentrations 160 times higher than expected in terrestrial rock. Most iridium in Earth sediments comes from extraterrestrial sources; in effect, iridium is the fallout from meteors as they disintegrate in the atmosphere. So the Alvarez team suggested that the greatly increased amounts of iridium supported the hypothesis that the great dinosaur die-off had been caused by immense quantities of dust injected into the atmosphere by the tremendous impact of an object from space.

Scientists everywhere leaped into the fray to explore, defend and criticize the Alvarez theory. In Snowbird, Utah, in October 1981, more than 100 scientists presented 70 scientific papers examining the Alvarez hypothesis and the effect of impacting bodies on Earth's evolution.

The impact theory won more support as scientists uncovered additional evidence. Researchers studying sections of Cretaceous-Tertiary boundary sediments at 50 sites around the world found abnormally high levels of iridium in 48 of them. Even more important, such iridium anomalies began cropping up in other layers, possibly marking other large extinctions.

Two University of Chicago paleontologists, David Raup and John Sepkoski, amassed a long list of all known families of marine animals (3,500 in all, comprising a quarter of a million species) and the dates of their appearance in and disappearance from the fossil record. With the aid of a computer they subjected the list to a statistical technique known as Fourier analysis, which detects patterns in seemingly random data. The study suggested a cycle of mass extinctions. The four most recent dyings occurred 11, 37, 66 and 91 million years ago — at intervals within a statistically acceptable range of 26 million years.

To account for the 26-million-year cycle, scientists turned to the cosmos. When Berkeley astronomer Richard Muller was shown a preprint of the Raup-Sepkoski report by his former teacher Luis Alvarez, he quickly calculated that the cause of the cycle might be a body traveling in an elongated orbit around the sun. In Muller's hypothesis, this body was either a dying star that had exhausted its nuclear fuel or a massive, Jupiter-like planet. As it neared the solar system every 26 million years or so, its gravitational mass dislodged a barrage of comets from their orbits in the Oort cloud, sending them plummeting toward the sun — and toward any planet in the way. Muller and colleagues Marc Davis and Piet Hut suggested several scholarly names for the body: Nemesis, the Greek goddess of retribution; Kali, the

This painting of a space probe approaching an asteroid illustrates one of a series of missions planned by NASA to study asteroids and comets for clues about the birth and early history of the solar system. Eventually, manned spacecraft might tow asteroids into Earth's orbit, where they could be mined for rare minerals.

154

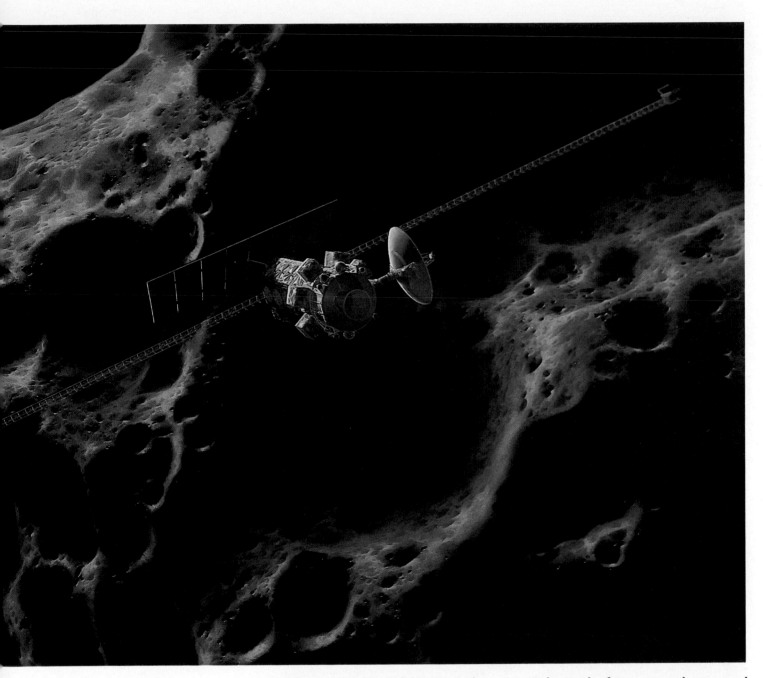

Hindu goddess of death; Indra, the Vedic god of storms and war; and George, the saint who slew the dragon. The press called it the Death Star.

By strict scientific criteria, the Alvarez theory and the more controversial hypotheses it inspired remain unproved. But they have already proved, if proof were needed, that the study of comets and asteroids is not merely a matter of academic curiosity — that anything and everything in Earth's cosmic environment may have a real effect on Earth's ecosystem and on the creatures who live and die there.

In extending their studies beyond the solar system, scientists have naturally confronted a host of problems, some predictable and some quite unexpected. In many cases, they have developed solutions by a combination of observation and rigorous logic. Other problems may prove insoluble.

A surprisingly difficult task has been to establish the physical limits of the solar system. Because the Oort-cloud comets are so small and so distant, there are no visible bodies on which to base revealing studies of the sun's diminishing gravitational pull. Not even the inner edge of the Oort cloud can be defined with any precision; it has been tentatively placed at 10 to 20 astronomical units, or 930 million to 1.86 billion miles from the sun. The outer edge of the Oort cloud has been calculated only in the roughest way. Astrogeologist Eugene Shoemaker estimates that the boundary lies 100,000 astronomical units from the sun — 9.3 trillion miles. In any case, the distances these regions involve are of an order far greater than those between the remote and widespread outer planets. As an index, Shoemaker cites the journey of the spacecraft *Pioneer 11,* which will take approximately 17 years from its launch to its arrival in the vicinity of Pluto. From there on, the spacecraft will be slowing down, but that will account for only a fraction of the time of its outward journey. Shoemaker estimates that Pioneer will take about a million years to travel from Pluto to the Oort cloud and "another 10 million years to finish the trip to the outer edge of the solar system."

In pressing their research beyond the solar system, scientists have exploited a formidable array of new and improved observational systems. In addition to orbiting telescopes and remote space probes, their technology includes sophisticated devices that have vastly increased the efficiency of older ground-based telescopes. Basic to many of the new systems are light detectors made of silicon (the material in grains of sand and computer

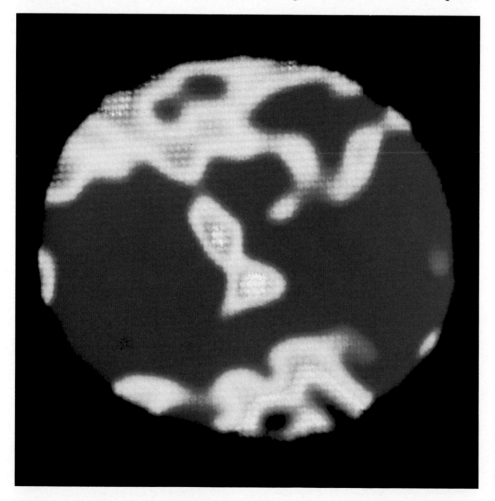

Betelgeuse, a star 500 times bigger than the sun, is 650 light years away — just a short trip in galactic terms. In this image, computer-assigned colors represent temperature differences across the star's surface. Orange indicates hotter regions, while cooler areas appear in blue.

chips) and known as CCDs, or Charge-Coupled Devices. The silicon detectors record light 100 times more efficiently than the most sensitive photographic film, and advanced CCDs can capture seven out of every 10 photons — the smallest particles of light — that fall upon them.

An older but no less important tool is the radio telescope, which mounts a large parabolic reflector — usually called a dish — to gather the radio waves emitted by distant bodies on various wavelengths. These telescopes, which came into common use in the 1950s, are particularly valuable because ground-based astronomical observation on radio wavelengths is not affected by atmospheric distortion, as it is on visual wavelengths. The range and capability of radio telescopes have been greatly extended by enlarging their dishes and by using several dishes in tandem on long base lines in order to get clearer composite images of remote objects. The biggest dish, in Arecibo, Puerto Rico, is 1,000 feet across, and the Very Large Array (VLA) of radio telescopes near Socorro, New Mexico, deploys 27 movable dishes, each 82 feet across, along 22 miles of railroad track.

Thanks to radio telescopes and other marvelous instruments, astronomers are peering far beyond the vague limits of the solar system. Exact observation has shown that the solar system lies in the lower part of a trailing arm of the Milky Way galaxy, a spiral galaxy with a span of about 100,000 light years, or nearly 600,000 trillion miles. At least 100 billion stars lie within the galaxy. For all that, the Milky Way is a galaxy of unexceptional size — among perhaps 100 billion other galaxies.

Powerful X-rays, shown here in blue, come from matter that — astronomers believe — is being torn away from a star and dragged at high speed into an invisible black hole, designated Cygnus X-1. Black holes are thought to be collapsed stars or other mass concentrations so dense that not even light can escape their incredible gravity. Cygnus X-1 was detected in 1971 by a telescope aboard a satellite orbiting Earth.

Just as gravity holds the solar system together, so gravity links the Milky Way to a local group of 30 other galaxies. The nearest major galaxy in the group is Andromeda, 2.2 million light years away. In turn the local group is tenuously linked by gravity to a supercluster of 2,500 galaxies centered on the immense Virgo galaxy more than 50 million light years away.

With the sophisticated instruments at their disposal, scientists have glimpsed strange phenomena that lie far beyond this supercluster, and that are difficult to perceive and understand even when they occur inside the Milky Way galaxy. Among these problematic objects are neutron stars, black holes and quasars.

The existence of neutron stars was postulated by Soviet physicist Lev Landau in 1932. His hypothesis, amplified by a supporting theory presented soon after, speculated that neutron stars are the remnants of massive stars that have exhausted their fuel and shattered their outer shells in supernova explosions. In the process, a star's remaining core is greatly compressed, and its atoms are crushed into smaller, denser neutrons — normally particles in the nuclei of atoms. Thus the neutron star is very small and very dense with a powerful magnetic field, and it radiates intense energy at various wavelengths. In its general outline this theory was later confirmed — though not for three decades and more.

The first neutron star was discovered in 1967 by a team of British radio astronomers at Cambridge University. Graduate student Jocelyn Bell and team leader Antony Hewish detected, in a constellation about 424 light years distant, a vigorous source of radio waves that emitted impulses with clocklike regularity at intervals of about one second. Because of these pulses, the object was called a pulsar.

As other pulsars were discovered, scientists concluded that they were rapidly spinning versions of the same phenomenon that Lev Landau and others had postulated three decades earlier. They deduced that the pulsars' magnetic poles lie at an angle to their axis of rotation, and thus they emit a ticklike impulse once or twice during each revolution. Pulsars have different rotation times, some completing a revolution in a fraction of a second and others taking several seconds. Neutron stars come in different sizes and densities. Typically, a neutron star about 5,000 light years away in the Crab Nebula is only 15 miles across but contains as much matter as the sun, which has a diameter 58,000 times greater.

Like neutron stars, the puzzling objects known as black holes were postulated before they were discovered — long before, in fact. Nearly 200 years ago, astronomer-mathematician Pierre Laplace postulated that if a celestial body with the sun's mass were only a few miles in diameter, its gravity would be so powerful that nothing — not even a ray of light — could escape from it. Later theorists proposed a likely process by which black holes could be created — and probably are. By actual observation, stars of ordinary size, such as the sun, become small, dense white dwarfs after they explode and collapse in their lengthy death throes. More massive stars — those with up to three times the mass of the sun — become smaller, denser neutron stars. So it stands to reason that supergiant stars, such as Betelgeuse, which is 20 times more massive than the sun, die away into still smaller and denser objects. Theoretically, the core collapses with terrific force and keeps on collapsing until it is compacted into a so-called singularity — a region that has no length, width or depth. In other words, the core disappears from the

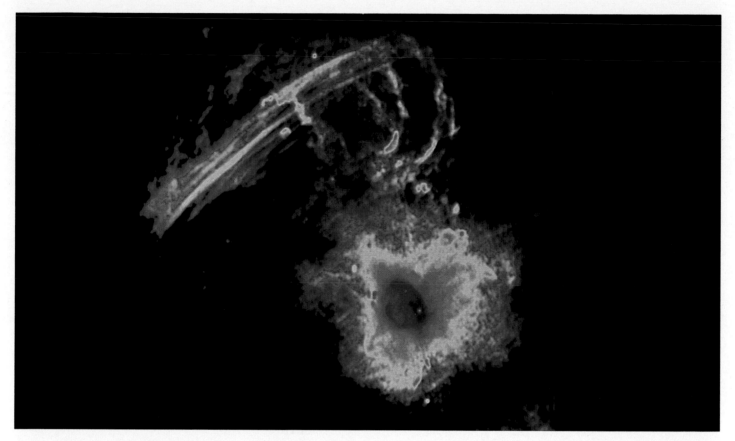

The mysterious core of the Milky Way galaxy and an immense arc of gaseous particles linked to it appear in false colors in this dazzling image based on radio-telescope data. Scientists believe that the arc, whose parallel filaments protrude at least 120 light years (about 600 trillion miles) from the galactic plane, is held in place by an enormous magnetic field generated in some unknown way at the center of the galaxy. The brightness of the particles is represented, in order of increasing intensity, by the colors green, blue, yellow and red.

observable universe and leaves an emptiness with a gravitational pull that is billions and even trillions of times more powerful than Earth's gravity.

Starting in the late 1960s, the first indications that black holes actually exist were detected by instruments aboard spacecraft. Though black holes are themselves invisible, their presence was revealed by tremendous radiations of energy from gas and dust, coming from nearby stars and attracted by the holes' voracious gravity. Moreover, a black hole's gravity visibly bends light rays in surrounding areas and literally warps space in much the same way that the impact of a falling lead ball deforms a sheet of rubber.

By the early 1980s, several hundred possible black holes had been located. But the study of the phenomenon is still in its infancy, with many questions unanswered except by pure speculation. One obvious question: What happens to all the matter that streams into a black hole's maw and disappears? One answer — sounding much like science fiction — is that the material travels through a tubelike aperture and emerges from a hypothetical white hole — into an entirely different universe.

If black holes are not the deepest mystery of the cosmos, quasars are. The first quasars were discovered in the early 1960s and were immediately recognized as perplexing anomalies. Radio astronomers detected them by their intense radio-wave emissions. Astronomers observing through optical telescopes saw them as stars of ordinary size but astonishing brightness, exceeding the luminosity of a whole galaxy of stars. But nothing about them accounted for their radio-wave activity or their brightness, which was even more baffling because they were apparently many billion light years away. On the ground that these phenomena might not be stars at all, they were called quasi-stellar radio sources, from which came the contraction quasar.

159

The mystery deepened with the discovery of more and more quasars — fully 1,500 by the early 1980s. It turned out that most of the quasars are not strong sources of radio waves, though all emit energy copiously on other wavelengths; therefore, purists renamed them quasi-stellar objects, or QSOs. Some quasars (still the name in general use) were found to have glowing extensions; these structures are thought to be jets of high-speed material shot from the active core of the quasars, but their cause and dynamics are unknown. And various apparent contradictions arose. The quasars' pattern of energy radiation varies over short periods, suggesting that the objects are on the small side. Yet it challenged credulity that a small object could be brighter than a galaxy.

Detailed studies of several quasars have highlighted two key aspects of the phenomenon: their tremendous distance and the great speed at which they are moving outward. MR-2251, one of the nearest quasars, is 1.2 billion light years distant. OQ172, one of the farthest, is 12 billion light years away, meaning that observed light from it started earthward only three to seven billion years after the Big Bang that gave birth to the universe. Probably because OQ172 is one of the farthest quasars, it is also one of the fastest-moving; incredibly, it is traveling outward at 167,400 miles per second, or 90 per cent of the speed of light. Even 3C273, one of the slower and nearer quasars, is receding at about 100,000 mph.

From what scientists have so far learned about quasars, they have formed some tentative opinions. Though quasars look like stars, it is only because of their great distance. Probably the typical quasar consists of a galaxy of a million or more stars clustered around an invisible object of tremendous mass, gravity and energy — most likely a black hole. Some scientists think quasars are exploding galaxies. Others suggest that the energy is radiated by clouds of gas and dust as they fall into the black hole.

Quasars and other perplexing phenomena have forced scientists to conjure with the very nature and extent of the universe. They have asked themselves many formidable questions. Will the quasars keep on traveling outward forever? Does the universe have enough matter, and hence gravity, to hold together, or will it keep expanding in all directions as it apparently has since the Big Bang? More and more, the answers to such questions, and the future of the universe, seem to be implicit in what happened during the Big Bang, the theoretic explosion that created the universe.

In a number of brilliant efforts to understand how the universe came into being, physicists and cosmologists are deducing their way backward in time toward the instant of creation. The theoreticians start with evidence that the Big Bang actually occurred. In 1965 at the Bell Laboratories in New Jersey, physicists Arno A. Penzias and Robert W. Wilson turned a radio antenna on the skies to search for microwave sources and found to their amazement that no matter where they looked, the intensity of microwave radiation never varied. It came from every part of space and uniformly corresponded to a temperature of $-454.8°$ F., just $5°$ F. above absolute zero, the actual temperature of space throughout the universe. Scientists interpret this background radiation as the faint afterglow of the Big Bang. Penzias and Wilson were awarded a Nobel Prize for their discovery.

For a second basic premise, the theoreticians have the seminal discovery made in the 1920s by astronomer Edwin P. Hubble. Using the powerful new 100-inch telescope atop Mount Wilson in California, Hubble found

The two galaxies NGC 4676A and B, moving through space 300 million light years away from Earth, pass so close to each other that the pull of gravity forms a connecting bridge of stars and other matter. In such encounters, both galaxies may lose matter not only at their point of contact but at their far ends; the disturbance of their internal gravities permits long tails of gas and dust to break free.

that the universe was receding in all directions. Hubble deduced this from studies of the so-called Doppler shift in the light from distant galaxies. Scientists refer to the spectra of galaxies as being "red-shifted," meaning that the absorption lines of various elements are displaced toward the red end of the spectrum — a sign that the star is moving away from the observer. Moreover, the speed of recession increases with distance. This phenomenon is known as Hubble's law. It is used by astronomers in calculating the scale of the universe and the speed of quasars in their outward journeys.

Scientists are in general agreement about the broad outlines of the Big Bang theory: Out of nothing, an explosion brought space, time and matter into existence and created an expanding universe. But some important questions remain. A principal problem is the total amount of mass in the cosmos, which will determine whether the universe will expand forever or whether the expansion will be slowed and finally reversed by gravitational forces. If there is enough mass, contraction and compression will prevail, and the universe could end in a so-called Big Crunch, which would obliterate all matter and all dimensions, 50 to 100 billion years hence.

The amount of mass required to hold the universe together is not very great; it works out to a density of scarcely one proton per 88 gallons of space — the capacity of a large bathtub. Yet astronomers estimate that the total mass of all the universe's galaxies and intergalactic dust clouds amounts to only about 1/40 the mass needed to provide enough gravity to halt the expansion of the universe. However, many scientists believe that there is no shortage of mass — that sufficient mass lies hidden in black holes or in supermassive but invisible clouds of gas around the galaxies.

Another problem with the Big Bang theory is understanding what occurred during the first fraction of a second after absolute Time Zero, when, according to mathematical extrapolations, the universe was heated to several trillion degrees — a million times hotter than the interior of the sun — and was unimaginably dense. Under such extraordinary conditions, normal physics no longer applies. As a result, scientists working to describe events at the instant of creation have devised a new set of rules and assumptions known as Grand Unified Theories — GUTs for short.

The various GUTs seek to embrace and reconcile, in a single elegant mathematical concept, all physical laws and accepted theories, from gravity to relativity to quantum mechanics. To do so, however, these theories postulate dozens of theoretic subatomic particles with strange names like X-bosons and gluons. To prove the actual existence of such fundamental particles, scientists normally test their theories in gigantic atom smashers or accelerators. Tests have already proved that certain particles deduced in the GUTs do in fact exist, and to confirm others bigger accelerators are being planned and constructed. But it may be impossible to build an accelerator big enough to prove the existence of every theoretic particle.

It is here that tomorrow's astronomical research will come to the rescue: Scientists may be able to test the Grand Unified Theories indirectly — by observing events in the farthest reaches of space to see if they follow predictions based on the initial conditions of the universe as described in the GUTs. Since ground-based telescopes already observe events occurring two thirds of the way back to the Big Bang, advanced satellite instruments, working beyond Earth's atmosphere, are expected to reveal much earlier events, such as the formation of the first galaxies out of the primordial

162

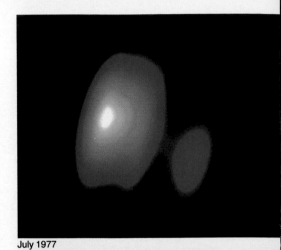

July 1977

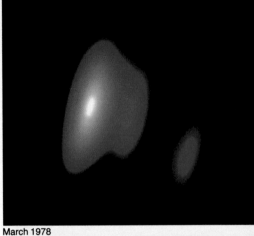

March 1978

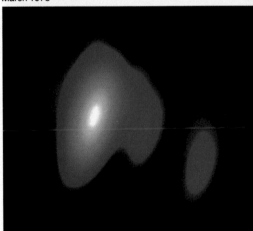

June 1979

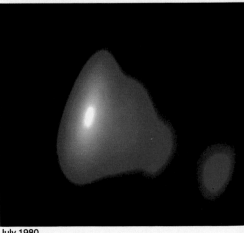

July 1980

matter that exploded into existence in the Big Bang. Properly interpreted, these observations could shed light on events in the very instant of creation.

To extend the reach of science farther into space and further back into time, NASA has set in motion practical plans for missions to be conducted during the remainder of the 20th Century and well into the 21st Century. The Astrophysics Division of NASA has devised experiments and missions concentrated in four broad fields of investigation: high-energy physics, ultraviolet and optical research, infrared and radio-wave research, and the study of relativity. NASA, with the help of a group of scientists called the Solar System Exploration Committee, has planned missions of "high scientific priority, moderate technological challenge and modest cost," among whose objectives is "the survey of resources available in near-Earth space in order to develop a scientific basis for future utilization of these resources."

The principal NASA missions are best known by the names and acronyms of the space vehicles that will carry them out. SOT, the Space Orbiting Telescope, will be followed by the Gamma Ray Observatory; by AXAF, the Advanced X-ray Astrophysics Facility; and by SIRTF, the Space Infrared Telescope Facility. These four observatories will scan the cosmos from the "long," or infrared, end of the electromagnetic spectrum to the intense short-wave rays. SIRTF will build on the discoveries of its fabulously successful precursor IRAS, probing the birth of stars and galaxies out of the cold primordial dust. SOT and another satellite, EUVE, the Extreme Ultraviolet Explorer, will look at stars in the extreme end of the ultraviolet range and also at visible wavelengths, developing new insight into the evolution of stars and galaxies just after they have formed. AXAF and the Gamma Ray Observatory will examine explosive events associated with the death of stars. Such explosions supply old star matter to be recycled into new stars.

NASA has also planned missions to advance knowledge of specific planets. These include the Venus Radar Mapper, the Mars Geoscience/Climatology Orbiter and the Galileo project, which will explore the Jupiter system with a planetary probe and an orbiter.

Astronomer Martin Harwit of Cornell University estimates that science has so far discovered a mere fraction — one fifth at most — of all the information to be learned about the universe. There is sound reason to expect that these and other new instruments will fill in much of the vast unpainted canvas. They will surely see more clearly than ever before the functions and the beauties of the solar system.

Science no less than poetry teaches that the solar system is a peaceful abode. Its center and controlling force, the star called the sun, provides the warmth and stability that has made possible life on Earth and the evolution of creative intelligence. Its spectacular planets, satellite worlds and mysterious small bodies exert a compelling attraction. They have drawn human beings to them, first in imagination and now in surrogate robot spacecraft, and one day will attract them in person.

There is nothing strange about the appeal of these celestial objects. Earth's human inhabitants are children of the cosmos. Their bodies contain minerals that were created from hydrogen and helium when primordial matter was recycled under tremendous heat and pressure to form the solar system. It is in the very nature of these beings to explore the cosmos, seeking to understand better where they came from. **Ω**

Quasar 3C 273, consisting of two closely spaced objects more than two billion light years distant, is shown expanding at speeds in excess of 30,000 miles per second in this series of four radio-telescope photographs taken in consecutive years. Quasars are enigmatic phenomena; little is known about them except that they are tremendously powerful sources of energy emissions. This one is 400 times more brilliant than the Milky Way galaxy.

RENEWING THE GALAXY

The recycling of matter, by which gas and dust from defunct stars formed the solar system 4.6 billion years ago, has become a familiar process to scientists, thanks to new tools of observation and data collection. Indeed all phases of the process, from the explosive death of giant stars to the birth of new stars, can be seen in four dazzling photographs of the Milky Way galaxy. The pictures, shown on these pages, were taken through an optical telescope by the noted David Malin, astronomer and research photographer at the Anglo-Australian Observatory in New South Wales.

Yet despite the lengthening range of their instruments, scientists suffered a particularly frustrating failure: Nowhere in the universe — nowhere in the 100 billion galaxies with at least 100 million stars each — were the researchers able to find convincing evidence of the existence of another solar system. But their search finally bore encouraging results in the early 1980s.

The long-awaited discovery, with the possibility it implies of finding life on other planets, was made by the orbiting vehicle IRAS (Infrared Astronomical Satellite) in 1983. IRAS's infrared scanning system, operating efficiently beyond Earth's filtering atmosphere, revealed an excess of infrared emission around approximately 40 stars within 75 light years of Earth. Some of these infrared sources can be interpreted as rings of dust and debris that could be birthing solar systems. Scientists now allow for the exciting possibility that planets may be forming in perhaps 20 clouds on Earth's own galactic doorstep.

The shattered outer layers of a dying star form a planetary nebula as they are blown into space, fluorescing brilliantly in ultraviolet radiation from the white dwarf that is the star's core. Originally, the white dwarf was a star with approximately the mass of the sun.

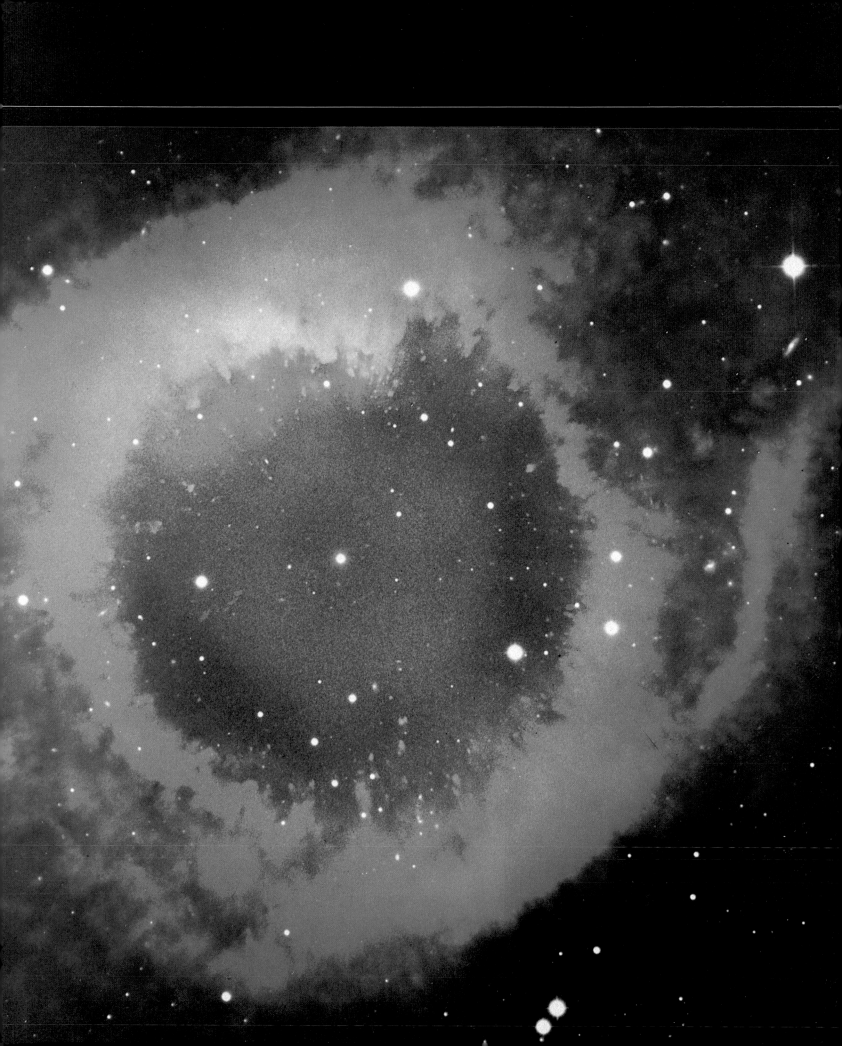

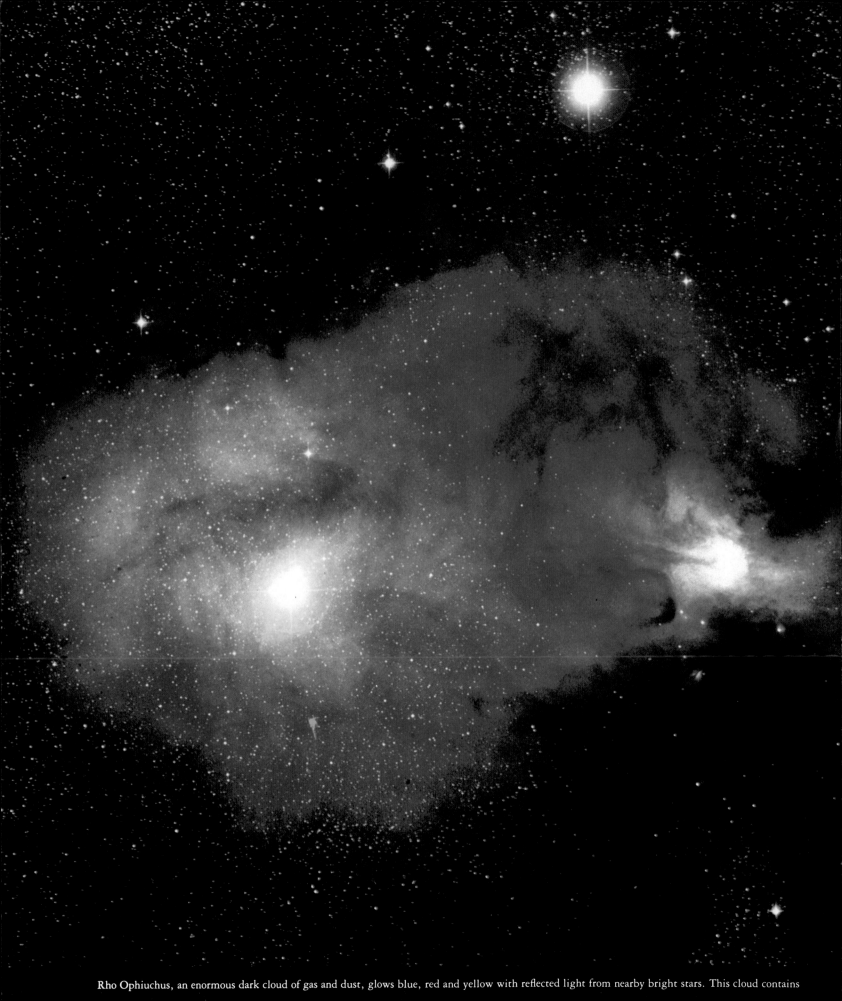

Rho Ophiuchus, an enormous dark cloud of gas and dust, glows blue, red and yellow with reflected light from nearby bright stars. This cloud contains

the raw materials for the formation of new stars; all that is needed is a cataclysmic event to stir the mixture.

Hot, swirling gases light the path of a shock wave racing through space at 600 miles a second. This violent aftermath of a supernova explosion, which

destroyed a massive star in the constellation Vela about 12,000 years ago, could compress a cloud of gas and dust and begin forming a new solar system.

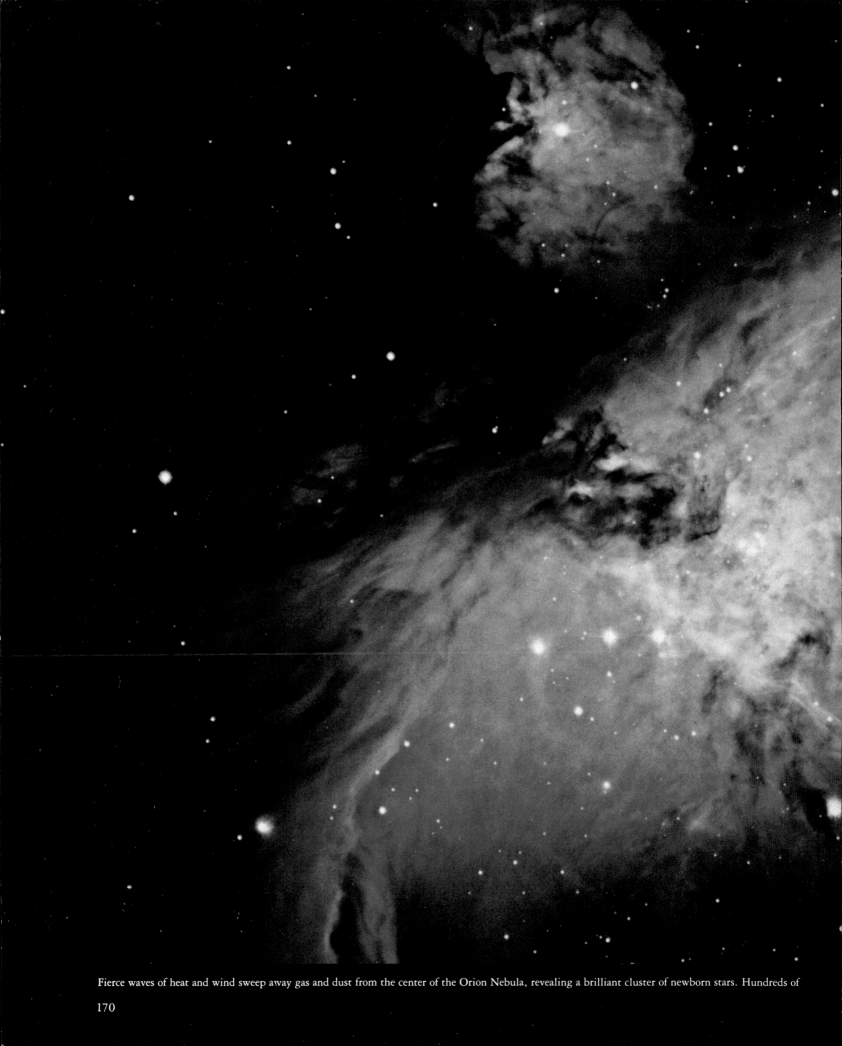

Fierce waves of heat and wind sweep away gas and dust from the center of the Orion Nebula, revealing a brilliant cluster of newborn stars. Hundreds of

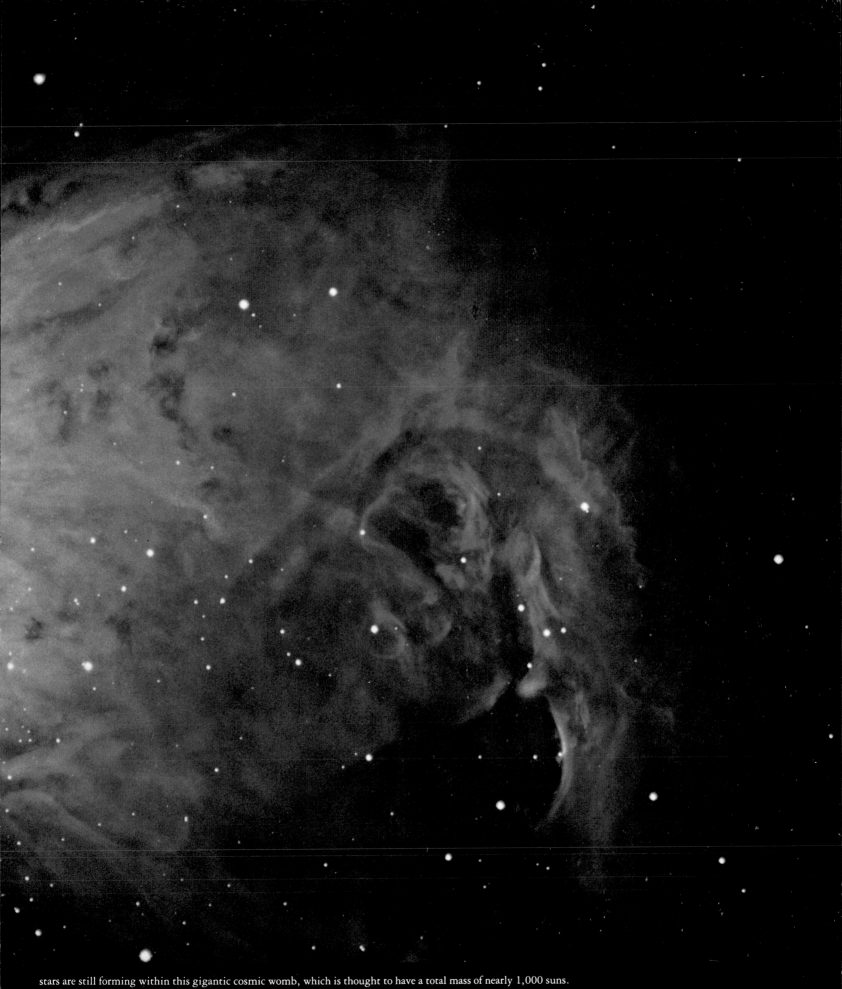

stars are still forming within this gigantic cosmic womb, which is thought to have a total mass of nearly 1,000 suns.

ACKNOWLEDGMENTS

For their help in the preparation of this book the editors wish to thank: **In France:** Pierre Soubrié, Société Astronomique de France, Paris. **In Great Britain:** Peter Hingley, Royal Astronomical Society, London. **In Italy:** Franca Principe, Istituto e Museo di Storia della Scienza, Florence. Luisa Ricciarini, Milan. Centro Internazionale A. Beltrame di storia dello spazio e del tempo, Vincenza. **In Japan:** Kazuaki Iwasaki, Osaka. **In the United States:** Arizona — Branch of Astrogeology, U.S. Geological Survey, Flagstaff: Raymond M. Batson, Eric Eliason, Alfred S. McEwen, Dr. Elliot C. Morris, Dr. Eugene M. Shoemaker, Dr. Paul Spudis, Joan D. Swann. Center for Meteorite Studies, Arizona State University, Tempe: Dr. Ronald Greeley and Dr. Carleton B. Moore. Robert Strom, Department of Planetary Sciences, University of Arizona, Tucson. California — Don Davis, Oakland. California Institute of Technology, Pasadena: Jane Beatrix and Barbara Wirick, Public Affairs Office; Dr. Andrew P. Ingersoll, Division of Geological and Planetary Science; Dr. Edward Stone, Division of Physics, Mathematics and Astronomy. Jet Propulsion Laboratory, Pasadena: Frank E. Bristow, Don Bane, Jurrie van der Woude, Office of Public Information; Stewart A. Collins; Leslie J. Pieri, Planetary Image Facility. District of Columbia — National Aeronautics and Space Administration: Dr. David Bohlin, Chief of Solar Physics; Joseph M. Boyce, Planetary Geology Program; Dr. Bevan M. French; Dr. Neil R. Sheeley Jr., Naval Research Laboratory. Maryland — Rob Wood, Annapolis. Dr. Norman Ness and Dr. Gerald Soffen, Goddard Space Flight Center, Greenbelt. L. Harper Pryor, Applied Research Corporation, Landover. Massachusetts — Dr. Fred Whipple, Smithsonian Astrophysics Observatory, Cambridge. New York — Dr. Rhodes W. Fairbridge, New York City. Dr. Tobias Owen, Department of Earth and Space Sciences, State University of New York, Stony Brook. Virginia — Mary Hurlbut, On The Line Graphics, Alexandria. William J. Hennessy Jr., Annandale. Walter Hilmers, Arlington. **In West Germany:** Dr. Erich Lamla, Sternwarte der Universitat, Bonn. Dr. Ernst Berninger and Gerhard Hartl, Deutsches Museum, Munich. Heidi Klein, Bildarchiv Preussischer Kulturbesitz, West Berlin.

The index was prepared by Barbara L. Klein.

BIBLIOGRAPHY

Books

Abbot, Charles Greeley, *The Sun and the Welfare of Man*. Smithsonian Institution Series, 1929.

Abell, George O., *Exploration of the Universe*. Saunders College Publishing, 1982.

Asimov, Isaac, *Asimov's Biographical Encyclopedia of Science and Technology*. Doubleday, 1982.

Beatty, J. Kelly, Brian O'Leary and Andrew Chaikin, eds., *The New Solar System*. Sky Publishing and Cambridge University Press, 1981.

Bergamini, David, and the Editors of Time-Life Books, *The Universe*. 2nd ed. Time-Life Books, 1977.

Bishop, Roy L., ed., *Observer's Handbook, 1984*. The Royal Astronomical Society of Canada, 1984.

Carr, Michael H., *The Surface of Mars*. Yale University Press, 1981.

Chapman, Clark R., *Planets of Rock and Ice*. Scribner's, 1982.

Cooper, Henry S. F., Jr., *A House in Space*. Holt, Rinehart and Winston, 1976.

Cortright, Edgar M., ed., *Apollo Expeditions to the Moon*. NASA SP-350, National Aeronautics and Space Administration, 1979.

Cronin, Vincent, *The View from Planet Earth*. William Morrow, 1981.

Davies, Merton E., et al., *Atlas of Mercury*. NASA SP-423, National Aeronautics and Space Administration, 1978.

Dunne, James A., and Eric Burgess, *The Voyage of Mariner 10*. NASA SP-424, National Aeronautics and Space Administration, 1978.

Eddy, John A., *A New Sun: The Solar Results from Skylab*. NASA SP-402, National Aeronautics and Space Administration, 1979.

Feldman, Anthony, *Space*. Facts on File, 1980.

Ferris, Timothy, *Galaxies*. Stewart, Tabori & Chang, 1982.

Fimmel, Richard O., Lawrence Colin and Eric Burgess, *Pioneer Venus*. NASA SP-461, National Aeronautics and Space Administration, 1983.

Frazier, Kendrick, *Our Turbulent Sun*. Prentice-Hall, 1982.

French, Bevan M., *The Moon Book*. Penguin Books, 1977.

Gibson, Edward G., *The Quiet Sun*. NASA SP-303, National Aeronautics and Space Administration, 1973.

Gillispie, Charles Coulston, ed., *Dictionary of Scientific Biography*. Scribner's, 1981.

Hartmann, William K.:
Astronomy: The Cosmic Journey. Wadsworth Publishing, 1978.
Moons and Planets. 2nd ed. Wadsworth Publishing, 1983.

Hartmann, William K., and Odell Raper, *The New Mars: The Discoveries of Mariner 9*. NASA SP-337, National Aeronautics and Space Administration, 1974.

Heath, Sir Thomas, *Aristarchus of Samos: The Ancient Copernicus*. Dover Publications, 1981.

Henbest, Nigel, *Mysteries of the Universe*. Van Nostrand Reinhold, 1981.

Hunt, Garry, and Patrick Moore, *Jupiter*. Rand McNally, 1981.

Hutchins, Robert Maynard, ed., *Ptolemy, Copernicus, Kepler (Great Books of the Western World*, Vol. 16). Encyclopaedia Britannica, 1952.

Jastrow, Robert, *Until the Sun Dies*. W. W. Norton, 1977.

Kaufmann, William J., III:
Black Holes and Warped Spacetime. W. H. Freeman, 1979.
Planets and Moons. W. H. Freeman, 1979.
Stars and Nebulas. W. H. Freeman, 1978.

King, Henry C., *The History of the Telescope*. Dover Publications, 1955.

Lockyer, Sir Norman, *The Sun's Place in Nature*. Macmillan, 1897.

Lodge, Sir Oliver, *Pioneers of Science and the Development of Their Scientific Theories*. Dover Publications, 1960.

Mars as Viewed by Mariner 9. NASA SP-329, National Aeronautics and Space Administration, 1974.

Meadows, A. J., *Early Solar Physics*. Pergamon Press, 1970.

Miller, Ron, and William K. Hartmann, *The Grand Tour: A Traveler's Guide to the Solar System*. Workman Publishing, 1981.

Mitton, Simon, ed., *The Cambridge Encyclopaedia of Astronomy*. Jonathan Cape, 1977.

Moore, Patrick:
The Atlas of the Universe. Rand McNally, 1970.
The Sun. W. W. Norton, 1968.
Watchers of the Stars: The Scientific Revolution. Michael Joseph, 1974.

Morrison, David, *Voyages to Saturn*. NASA SP-451, National Aeronautics and Space Administration, 1982.

Morrison, David, and Jane Samz, *Voyage to Jupiter*. NASA SP-439, National Aeronautics and Space Administration, 1980.

Nicolson, Iain, *The Sun*. Rand McNally, 1982.

Noyes, Robert W., *The Sun, Our Star*. Harvard University Press, 1982.

Pasachoff, Jay M., and Marc L. Kutner, *University Astronomy*. W. B. Saunders, 1978.

Povenmire, Harold R., *Fireballs, Meteors and Meteorites*. JSB Enterprises, 1980.

Ridpath, Ian, *Stars and Planets*. Hamlyn, 1978.

Ridpath, Ian, ed., *The Illustrated Encyclopedia of Astronomy and Space*. Revised ed. Thomas Y. Crowell, 1979.

Ronan, Colin A., *Deep Space*. Macmillan, 1982.

Sagan, Carl, *Cosmos*. Random House, 1980.

Seyfert, Carl K., and Leslie A. Sirkin, *Earth History and Plate Tectonics*. Harper and Row, 1979.

Silk, Joseph, *The Big Bang: The Creation and Evolution of the Universe*. W. H. Freeman, 1980.

Smithsonian Exposition Books, *Fire of Life*. W. W. Norton, 1981.

Stephen, Sir Leslie, and Sir Sidney Lee, eds., *Brown-Chaloner (The Dictionary of National Biography*, Vol. 3). Oxford University Press, 1950.

Tauber, Gerald E., *Man's View of the Universe*. Crown, 1979.

Tombaugh, Clyde W., and Patrick Moore, *Out of the Darkness: The Planet Pluto*. Stackpole Books, 1980.

Turner, Gerard L'Estrange, *Antique Scientific Instruments*. Blandford Press, 1980.

White, Oran R., ed., *The Solar Output and Its Variation*. Colorado Associated Press, 1977.

Wilkening, Laurel L., ed., *Comets*. The University of Arizona Press, 1982.

Wood, John A., *The Solar System*. Prentice-Hall, 1979.

Young, C. A., *The Sun*. D. Appleton and Company, 1895.

Periodicals

Alvarez, Luis W., et al., "Extraterrestrial Cause for the Cretaceous-Tertiary Extinction." *Science*, June 6, 1980.

Alvarez, Walter, et al., "Impact Theory of Mass Extinctions and the Invertebrate Fossil Record." *Science*, March 16, 1984.

Anderson, Charlene M., "Asteroid Project Discovers Ten New Asteroids." *The Planetary Report*, May/June 1984.

Anderson, Don L., "The Earth as a Planet: Paradigms and Paradoxes." *Science*, January 27, 1984.

Astronomy, March 1984.

Berry, Richard, "Mysterious Pluto." *Astronomy*, July 1980.

Burnham, Robert, "The Saturnian Moons." *Astronomy*, December 1981.

Chaiken, Andrew, "Target: Tunguska." *Sky & Telescope*, January 1984.

Chaisson, Eric, "Journey to the Center of the Galaxy." *Astronomy*, August 1980.

Chapman, Clark R., "The Nature of Asteroids." *Scientific American*, January 1975.

Ciaccio, Edward J.:
"Atmospheres." *Astronomy*, May 1984.
"Celestial Debris." *Astronomy*, May 1983.

Cloud, Preston, "The Biosphere." *Scientific American*, September 1983.

"Cosmic Winter." *Discover*, May 1984.

Couper, Heather, "Journey to the Center of the Galaxy." *New Scientist*, April 26, 1984.

Cowley, S.W.H., "Jupiter's Magnetosphere." *Nature*, October 30, 1980.

"Cycles of Extinction." *Discover*, May 1984.

"Darkness at Noon." *Discover*, August 1983.

"Death Star." *Science News*, April 21, 1984.

Di Cicco, Dennis, "Wethersfield Meteorite: The Odds were Astronomical." *Sky & Telescope*, February 1983.

Drake, Stillman, and Charles T. Kowal, "Galileo's Sighting of Neptune." *Scientific American*, December 1980.

Driscoll, Everly, "Mariner Views a Dynamic, Volcanic Mars." *Science News*, February 12, 1972.

"The Dry, Cold, Rocky Ground of Mars." *Science News*, July 24, 1976.

Eberhart, Jonathan, "The Last Viking." *The Planetary Report*, September/October 1983.

Edberg, Stephen J., "Halley Watch '86." *Astronomy*, March 1983.

Ferris, Timothy, "Return of the Death Star." *Science Digest*, July 1984.

"50 Years after Clyde Tombaugh Found Pluto, His Career Is Still Looking Up." *People*, May 12, 1980.

Frazier, Kendrick:
"Our Flickering Sun: Fluctuating Solar Flux." *Science News*, May 1, 1982.

"A Planet beyond Pluto." *Mosaic*, September/October 1981.

"The Frigid World of IRAS-Parts I and II." *Sky & Telescope*, January and February 1984.

Ganapathy, R., "The Tunguska Explosion of 1908: Discovery of Meteoritic Debris near the Explosion Site and at the South Pole." *Science*, June 10, 1983.

Getts, Judy A., "Decoding the Hertzsprung-Russell Diagram." *Astronomy*, October 1983.

" 'Golden Age' of Astronomy Peers to Edge of Universe." *The New York Times*, May 8, 1984.

Gore, Rick, "What Voyager Saw: Jupiter's Dazzling Realm." *National Geographic*, January 1980.

"Great Balls of Fire." *Discover*, December 1981.

"The Great Dyings." *Discover*, May 1984.

Guth, Alan H., and Paul J. Steinhardt, "The Inflationary Universe." *Scientific American*, May 1984.

"In Orbit around Mars: Into the Second Week." *Science News*, November 27, 1971.

Johnson, Torrence V., and Laurence A. Soderblom, "Io." *Scientific American*, December 1983.

Kresák, L., "The Tunguska Object: A Fragment of Comet Encke?" *Bulletin of the Astronomical Institutes of Czechoslovakia*, Vol. 29, No. 3, 1978.

Lunine, Jonathan I., "Mucky Seas and Hazy Skies: An Ethane Ocean on Titan?" *The Planetary Report*, November/December 1983.

Malin, David:
"The Colors of Deep Space." *Astronomy*, March 1980.
"The Orion Nebulae in Color." *Sky & Telescope*, November 1981.

Mammana, Dennis L., "Astronomers Are Surprised to Find that Distant Uranus Is the Second Ringed Planet in the Solar System." *Smithsonian*, December 1978.

Maran, Stephen P.:
"Inside Jupiter's Rings." *Natural History*, August 1982.
"What Struck Tunguska?" *Natural History*, February 1984.

Marov, M. Y., "Highlights of the Venera Missions to Venus." *The Planetary Report*, November/December 1982.

Merritt, J. I., "Birth and Death of the Universe." *University*, Winter 1981.

Morrison, David, "Four New Worlds." *Astronomy*, September 1980.

North, J. D., "The Astrolabe." *Scientific American*, January 1974.

Nozette, Stewart, "Global Altimetry Map of Venus." *The Planetary Report*, February/March 1981.

Oberg, James E., "Tunguska Collision with a Comet." *Astronomy*, December 1977.

O'Toole, Thomas, "Mystery Heavenly Body Discovered." *The Washington Post*, December 30, 1983.

Owen, Tobias:
"The Evolution of Titan's Atmosphere."

The Planetary Report, November/December 1983.

"The Largest Moon of Saturn, Observed by Voyager 1 and Voyager 2, May Have Oceans of Methane." *Scientific American*, February 1982.

Parker, E. N., "Magnetic Fields in the Cosmos." *Scientific American*, August 1983.

"Periodic Impacts and Extinctions Reported." *Science*, March 1984.

Pieri, David, "The Ancient Rivers of Mars." *The Planetary Report*, January/February 1983.

The Planetary Report, July/August 1983.

Reddy, Francis, "Backtracking the Comets." *Astronomy*, August 1982.

Rubin, Alan E., "Chondrites and the Early Solar System." *Astronomy*, February 1984.

Schwartzenburg, Dewey, "The Great Red Spot." *Astronomy*, July 1980.

Sekanina, Z., "The Tunguska Event: No Cometary Signature in Evidence." *The Astronomical Journal*, September 1983.

Soderblom, Laurence A., "The Galilean Moons of Jupiter." *Scientific American*, January 1980.

Soderblom, Laurence A., and Torrence V. Johnson, "The Moons of Saturn." *Scientific American*, January 1982.

Squyres, Steven W.:
"Ganymede and Callisto." *American Scientist*, January/February 1983.
"The Solar System's Other Ocean." *The Planetary Report*, May/June 1983.

Strand, Linda Joan, "The Star Tar in the Jupiter Jars." *Astronomy*, June 1984.

Sullivan, Walter, "Cause of 1908 Blast over Siberia Remains Elusive." *The New York Times*, July 1, 1983.

Trefil, James, "How the Universe Began." *Smithsonian*, May 1983.

Wagner, Jeffrey K., "The Sources of Meteorites." *Astronomy*, February 1984.

Weaver, Kenneth F., and James P. Blair, "The Incredible Universe." *National Geographic*, May 1974.

Other Publications

Aurora Borealis. Alaska Geographic Society, 1979.

Images of Mars: The Viking Extended Mission. NASA SP-444, National Aeronautics and Space Administration, 1980.

Planetary Exploration through Year 2000. Part 1 of a report by the Solar System Exploration Committee of the NASA Advisory Council. U.S. Government Printing Office, 1983.

The Planets. Readings from *Scientific American*. W. H. Freeman, 1983.

Pocket Statistics. National Aeronautics and Space Administration, 1979.

Viking: The Exploration of Mars. NASA SP-208, National Aeronautics and Space Administration, no date.

The Voyager Flights to Jupiter and Saturn. JPL 400-148, National Aeronautics and Space Administration, 1982.

PICTURE CREDITS

INDEX